Integrated Science

Book 3

Wreake Valley School Science Department

Syston   Leicestershire.

General Editors: Dirk Tinbergen and Phil Thorburn

Authors

Dave Barlex
Roy Bright
Dan Conboy
Sue Cooke
George Goldby

Maurice Kyle
Barbara Matthieson
Brian Sherwood
Phil Thorburn
Dirk Tinbergen

Edward Arnold

© Wreake Valley School Science Department 1976

First published 1976 by Edward Arnold (Publishers) Ltd
25 Hill Street, London W1X 8LL

ISBN 0 7131 0043 5

All Rights Reserved. No part of this publication may be
reproduced, stored in a retrieval system, or transmitted
in any form or by any means, electronic, mechanical,
photo-copying, recording or otherwise, without the
prior permission of Edward Arnold (Publishers) Ltd.

Layout and illustrations by
Reproduction Drawings Ltd.

Printed by Butler and Tanner Ltd,
Frome and London

Preface

This 'curriculum' in science for students between ages 11 and 14 is prepared by a group of working teachers who have thought and worked hard at how to engage beginners in that form of thinking called science. It is neither revolutionary in conception nor particularly innovative with respect to its materials, apparatus, or exercises — but I do not intend that in a negative sense. Rather, what strikes me about this curriculum is that it has built upon points of view and techniques that have been growing steadily in science teaching during the past decade in Britain, America, and Europe. The authors have been ingenious about the observations and experiments to which students may be introduced, bearing in mind not only the intrinsic interests they evoke, but also the centrality of the scientific ideas to which these interests can lead. The spirit of Nuffield in British science teaching, of the Elementary Science Study in America, and of a variety of other curriculum efforts permeate the venture. Yet, for all that, there is an admirable common sense in the approach. Let me say a word about that, for it may help teachers to a better sense of how such a curriculum as this one can be used.

The first thing about it is that it is not a 'finished' curriculum in the sense of imposing a fixed structure on the materials and ideas it encompasses. The authors rightly say that there are a variety of 'methods' that can be used in teaching it. It is *not* teacher-proof — as if anything *could* be teacher-proof. Rather, it represents the result of the wisdom and hard work of a group of working teachers who tried to adapt modern ideas of science teaching to their own class-room needs. The clever student can go through the exercises proposed with a minimum of help from the teacher. The slower student is well advised about the points at which he may need some consultation. The teacher who works with these materials will not find the 'line' laid down as to how to proceed with different kinds of students. *He* will have to work as well as the students. For the objective of the laboratory exercises is to open up the student's curiosity about phenomena he is bound to encounter in carrying out observations, measurements, and experiments. This curriculum is an invitation by teachers to other teachers to help students make sense of problems in science, some of which seem obvious but are not, others of which seem impenetrable but are not. It does this by creating situations that invite the student to think science rather than thinking *about* science. As such, it is no more fool-proof than science is fool-proof. For science, after all, is a way of thinking and of making one's thoughts subject to tests on nature. If such science teaching succeeds, it does so by making the student more independent in his thinking as a scientist, freer to pursue his thoughts where they may lead him by making him more discerning in eliminating his own errors and follies.

Any teacher or group of teachers who attempt to teach a course like this one will end up not with somebody else's new curriculum, but with their own. Like Nuffield science for older students, this course invites teachers to modify and to create approaches of their own — and they will have to, for the clever and the slow student will approach things differently, the more formally minded and the more intuitive will have different styles, and the more independent and more clinging will pose different demands on the teacher. Indeed, the lessons of the last decade all point to the endlessness of curriculum making. What this book provides is the stimulus, an ingenious and well conceived set of exercises that will challenge not only the student but the teacher. In this sense, it is as much an effort in teacher development as it is in curriculum making.

Jerome Bruner

Watts Professor of Psychology
University of Oxford.

16 February, 1975

Introduction

This course

*has been written by practising teachers, in a comprehensive school.

*has been tested with mixed ability groups.

*is an integrated science course for 11 - 14 year olds.

*has step by step instructions using very simple language.

*has special 'Extension Work' for more able pupils.

How was each chapter written?

A team of 4, always including at least 1 biologist, 1 physicist and 1 chemist, were involved in the planning and writing of each chapter. After several draft stages, the chapter was tested. It was then modified by one of the team, together with two overall editors (Phil Thorburn and Dirk Tinbergen).

Has the course been tested?

The techniques used in the course have been used by several of the authors in their various schools before coming to Wreake Valley. The whole course has been tested with mixed ability groups, before final modification.

What age range does it cater for?

Years 1 -3 of the Secondary School. But teachers may regard it as suitable for Middle Schools.

Does it lead into C.S.E. and 'O' level courses?

Yes. The authors teach Nuffield courses in years 4 and 5.

What sort of school is the course aimed at?

We have written the course so that it can be used by mixed ability groups in laboratories equipped with standard apparatus. However, it is equally suitable for streamed classes.

What about the more able?

Some testing extension work has been included. Teachers can always provide more if they feel it is appropriate.

How does the non-reader manage?

Any course revolving around written instructions has to face this problem. Pupils do help each other to a certain extent. The course has been written in a style that pupils can understand and so can follow instructions without difficulty. This enables the teacher to be freed to give help where it is most needed.

Is the extension work for the most able pupils?

By and large, yes. We have included only a limited amount of extension work, which tends to be of a more difficult nature. We hope that some teachers will want to write their own extra extension work, for a wider ability range.

Does the course need special equipment?

As a practical course, it relies heavily on standard equipment. However, it has been designed to avoid using equipment not readily available to most schools.

Can the physicist cope with the biology?

In the testing of the course, the experience has been that, by having to teach biology topics, the physicist has spent much more time talking about the work with his colleagues than is usual. This can only be good. No real problems are then encountered.

How many lab staff are required?

The course has been tested in a school where 10 full-time science teachers relied on 2 lab. technicians.

How hard do I have to work?

The apparatus lists, and the detailed instructions given to pupils means that they can get on with their work, without putting undue pressure on the teacher.

What's the point of the equipment lists?

Each equipment list states what is required by 1 working group (the teacher decides how big this is). How many sets of apparatus are required depends on how the course is used — i.e. all pupils doing the same thing at the same time, or pupils working at their own rate.

Can it be taught by traditional methods?

The course can be taught in any way the teacher feels suitable. There is no right or wrong way.

Contents

Preface   iii

Introduction   iv

1. **Let there be light . . .**   1
   Unit 1 Light and darkness
   Unit 2 Bouncing light
   Unit 3 Bending light
   Unit 4 Mystery bending tricks
   Unit 5 Making images
   Unit 6 A Magnifying glass
   Unit 7 Making an image bigger
   Unit 8 Colours from white
   Unit 9 Mixing colours

2. **. . . And there was light**   10
   Unit 1 Better effects with prisms
   Unit 2 More about images
   Unit 3 How a lens forms an image
   Unit 4 Finding out why the image is upside-down
   Unit 5 A model telescope on the table-top
   Unit 6 Another virtual image
   Unit 7 More about reflection

3. **Camera and eye**   18
   Unit 1 Making a pinhole camera
   Unit 2 Letting more light in
   Unit 3 Trying another idea
   Unit 4 Making a lens camera
   Unit 5 Eye dissection
   Unit 6 Eye protection
   Unit 7 What are eyes for?

   Extension Work   25
   1 Illusions
   2 Eye positions

4. **Forces**   28
   Unit 1 What is a force
   Unit 2 Forces exhibition
       Exhibit 1 Forces on magnets
          2 Electrical forces
          3 Friction forces
          4 Spinning forces
          5 Elastic forces
          6 Gravity forces
   Unit 3 Balanced and unbalanced forces
   Unit 4 Why is it difficult to get things moving?
   Unit 5 What do we mean by pressure?

   Extension Work   39
   1 The model suspension bridge
   2 Testing the bridge
   3 How to work out pressure

5. **To the moon**   44
   Unit 1 Getting away from it all
   Unit 2 Out in space
   Unit 3 The moon's surface
   Unit 4 Some facts

   Extension Work   47
   Moon questions

6. **Making things work**   48
   Unit 1 Motors and engines
   Unit 2 The Human Engine
   Unit 3 How much fuel?

   Extension Work   52
   The electric motor — another worker

7. **Movement**   53
   Unit 1 What gets things moving?
   Unit 2 Falling objects
   Unit 3 The accurate measurement of motion

   Extension Work   58
   1 Acceleration graphs
   2 Looking at collisions

8. **Breaking up is hard to do**   60
   Unit 1 Heating some substances that are found naturally in our world
   Unit 2 Heating some purified substances in test tubes
   Unit 3 Different kinds of reaction
   Unit 4 Breaking up a residue

   Extension Work   69
   1 Oxygen
   2 Carbon dioxide

9. **Atoms**   70
   Unit 1 Atoms and elements
   Unit 2 Elements and compounds

10. **Elementary my dear compound**   75
    Unit 1 Making a compound with magnesium and oxygen
    Unit 2 Comparing properties
    Unit 3 Making a compound with carbon and oxygen

Extension work   78
1 Flame tests
2 Another test for metals
3 Examining oxides of metals and non-metals

11. Reactivity   81
Unit 1 What is reactivity?
Unit 2 Comparing the reactivity of metals with acid
Unit 3 Where have the metals gone?
Unit 4 Making copper join with chlorine
Unit 5 Can we get the metals back from the solutions?

Extension Work   87
1 Using a blow pipe
2 The blast furnace — making iron
3 Methods of extracting metals

12. Rates of reaction   89
Unit 1 The effect of heat on a reaction
Unit 2 Reacting zinc with acid at different temperatures
Unit 3 Catalysts: another way to speed up a reaction

Extension Work   92
1 How does concentration affect the speed of a reaction
2 Reacting acid with marble chips of different sizes

13. Get a move on   94
Unit 1 Evaporation: a reminder
Unit 2 Freezing
Unit 3 How particles move in a solid
Unit 4 Melting points
Unit 5 How atoms are joined together
Unit 6 Testing solids

Extension Work   101
1 Electrolysis
2 Metals
3 Growing crystals

14. Merging molecules   103
Unit 1 Predicting what will happen when sulphur is heated
Unit 2 Heating sulphur
Unit 3 Making Nylon: a chemical rope trick
Unit 4 Making Nylon — an experiment
Unit 5 Poly compounds
Unit 6 Breaking down and repolymerising Perspex
Unit 7 The polymer starch
Unit 8 A catalyst again

Extension Work   115
1 Proteins
2 Coal
3 Oil

15. Grub, guts and spit   119
Unit 1 Food tests
Unit 2 The model body
Unit 3 Imitation food
Unit 4 Real food
Unit 5 Useless food?
Unit 6 What do we do with starch?

Extension work   126
1 Digest your eggs
2 Acid guts
3 Indigestion
4 Saliva versus starch

Equipment List   128

1

Let There Be Light...

Part A
Looking at light

Unit 1
Light and darkness

■ What to do

(a) Collect:
a light source
a small screen and stand
a low voltage supply
a small piece of card.

(b) Connect your light source to the low voltage supply. Ask your teacher to check your circuit.

(c) Set up the apparatus as shown below.

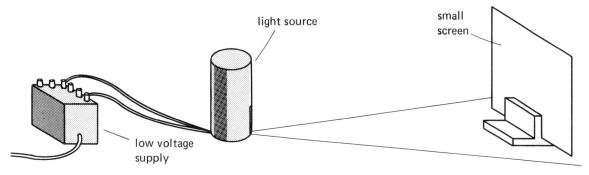

(d) Cut out a shape in the middle of the piece of card. Here are some ideas for the shape:

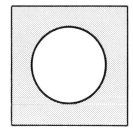 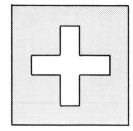  

(e) Hold the card between the light source and the screen, just in front of the screen:

▲ **What to write**

1. Why wasn't the light at B (in the diagram above) as bright (intense) as at A?
2. Was it completely black at B? Why not?
3. Was the size of the patch of *brighter* light (A)
    i. about the same as your cut-out?
    ii. larger than your cut-out?
    iii. smaller than your cut-out?
4. Was the patch clear, or blurred?

■ **What to do**

(f) Find out how to make the bright patch larger than the cut-out and more blurred than before.

▲ **What to write**

5. How did you make the patch larger?
6. What did you notice about the edge of the patch of light?
7. Look at the drawing below and notice how the beam of light spreads out. Then make a similar drawing, showing how light spreads out through *your* cut-out.
8. You know that the light is passing from the light source to the screen. Why can't you see the light at X (see diagram)?

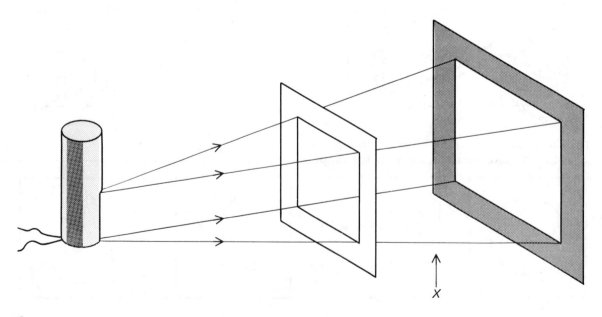

Unit 2
Bouncing light

★ **Information**

You may have realised that light travels out from a light source in straight lines.
Let's try changing its direction.

■ **What to do**

> (a) Fetch a light source
> a low voltage supply
> a flat (plane) mirror
> a large sheet of white paper.

(b) Put the sheet of white paper flat on the table, and shine a beam of light along it, like this:

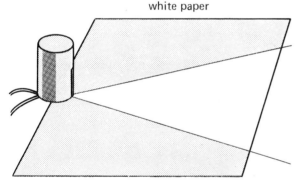

(c) Hold the mirror in the beam of light, at an angle to it, and notice what happens when the light hits it.

▲ **What to write**

1. When the light hits the mirror, does it stop there, or bounce off it?

2. Look into the mirror so that you can see the light source. Where does the light from the bulb go to after hitting the mirror?

★ **Information**

When light bounces off things, we say it is being *reflected*.

We only see things because they are either glowing themselves, like a candle flame or a lamp, or light is being reflected off them.

▲ **What to write**

3. Make a list of things which can produce light.

4. You see this page because light is bouncing off the page into your eye. Where does the light come from?

5. You can see the Moon at night because light is reflected off it, into your eye. Where does this light come from?

6. Which reflects light better, a black surface or a white surface?

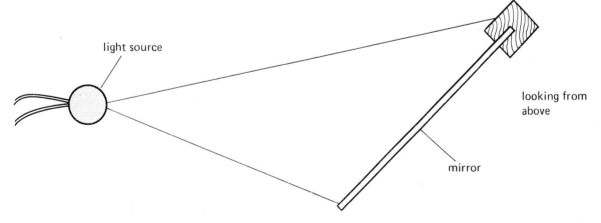

Unit 3
Bending light

▲ **What to write**

(a) Look at the smoke-box demonstration and notice that the beam of light is being bent by the glass lens

▲ **What to write**

1. What happens to the beam of light as it passes through the glass lens?

2. Whereabouts in the smoke-box is the light the brightest, at A, B, or C? (See diagram)

★ **Information**

We have found that we can change the direction of light by either

   i. making it bounce off things. This is called *Reflection*.

   ii. making it *bend* as it goes from air into glass, or from glass into air. This is called *Refraction*.

▲ **What to write**

3. Explain the two ways by which we can change the direction of a beam of light.

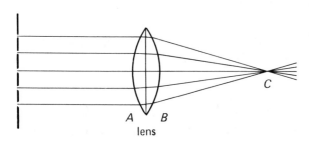

Unit 4
Mystery bending tricks

★ **Information**

Refraction can also occur when we use water instead of glass.

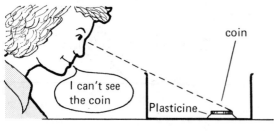

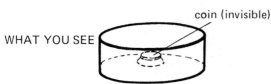

■ **What to do**

(a) Set up the apparatus as shown in the diagram

(b) Make sure you can *just not* see the coin.

(c) Now ask someone else to pour water *very slowly* into the container.
Keep your head still and your eyes open.

▲ **What to write**

1. Write down what you could see.

(d) Put a pencil into the water, as shown in the diagram.

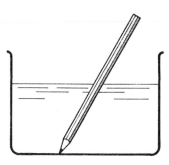

(e) Notice that your pencil seems to have bent? (Can you straighten it again?)

(f) Try this trick next time you have a bath:

Sit with your foot as far away from you as possible, just under the surface of the water. Look at the size of your toes as you slowly raise your foot out of the water. (The water must be clean.)

Part B
Making a telescope

Unit 5
Making images

■ **What to do**

(a) Fetch a round (spherical) lens and miniature screen. The lens looks something like this:

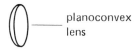
planoconvex lens

(b) Find yourself a place in the lab where there is a window, or other bright light source.

(c) Stand several metres from the window, with your back to it, and hold up your miniature screen at arm's length.

(d) Hold the lens upright in front of the screen, about 30 cm from it.

(e) Holding the screen still, move the lens towards it, until you see a clear, sharp picture of the window on the screen.

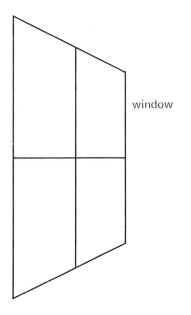

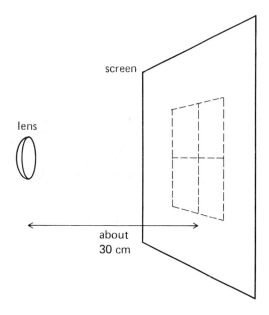

▲ What to write

1. What do you notice about the picture on the paper:

    Is it coloured?
    Is it clear or blurred?
    Is it smaller or larger than the real thing?
    Is it the right way up?

2. Can you change the distance between the lens and the screen and still get a clear picture?

3. Ask a friend to hold black paper to shield the screen from stray light. Does this improve the picture?

★ Information

The "picture" you get with the lens is called an *image*. If lenses didn't form images, we would not have spectacles, microscopes, telescopes and other *optical instruments*.

▲ What to write

4. Write down the names of three other optical instruments.

★ Information

If the object, of which you are making an image, is several metres away or more, the distance between the lens and the image is called the *focal length* of the lens.

▲ What to write

5. How far is your clear image from the lens?

6. What, therefore, is the focal length of your lens?

**Unit 6
A Magnifying glass**

★ Information

In Unit 1, you discovered that

   i. a lens can make a small image

   ii. the image is upside-down.

We can find out very easily what else a lens will do.

■ What to do

(a) Put this book on the table in front of you, and hold the lens flat on the book.

(b) Look into the lens, and move it slowly towards your eyes.

▲ What to write

1. What do you notice as you move the lens towards your eyes?

2. Is it upside-down?

3. As you bring the lens closer and closer to your eyes, does what you see continue to get larger and larger?

4. If not, say what happens to it.

Unit 7
Making an image bigger

■ What to do

(a) Collect:
a ½-metre ruler
a piece of greaseproof paper
some pieces of Plasticine
a + 7D lens

(b) Make a lens stand upright at one end of the ruler using Plasticine:

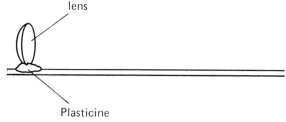

(c) Point the lens on the ruler at something far away and bright. Look along the ruler:

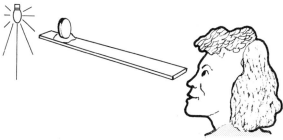

(d) Position the greaseproof paper between your eye and the lens, so that you get a clear image of the bright light on the paper. (You can see the image through the paper from where you are looking.)

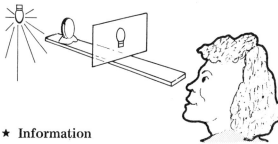

★ Information

Now that you have got an image on the paper, like we got in Unit 1, let's try to use a magnifying glass to make it bigger!

▲ What to write

1. What do you think this second lens will do to the image?

■ What to do

(e) Get a friend to position this other lens between the paper and your eye, so that it magnifies the image as much as possible:

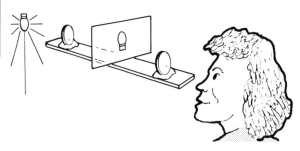

(f) Position the second lens with Plasticine.

(g) Take the paper away and keep looking along the ruler at the bright object.

▲ What to write

2. What did the image on the greaseproof screen look like?

3. What did the image look like when seen through the magnifying glass?

★ Information

Maybe we could magnify the image even more . . .

■ What to do

(h) Now fetch one of the very curved lenses. These are *stronger* magnifiers than the other lens you used.

(i) Replace the second lens with the very curved one. Position it to enlarge the image as much as possible.

(j) Now use your instrument as a telescope to look around the lab. (You may notice some colours round the edges of the images you see.)

▲ What to write

4. Did you magnify the image more using the very curved lens?

5. Why, do you think, is a real telescope enclosed in a tube?

Part C
Prisms and colours

Unit 8
Colours from white

■ What to do

(a) Collect:
a small 60° prism
a light source
a low voltage supply
a screen
a large sheet of white paper.

(b) Lay the sheet of paper flat on the table.

(c) Connect up the light source to a low voltage supply to make a beam of light across the paper. Ask your teacher for help if this seems difficult.

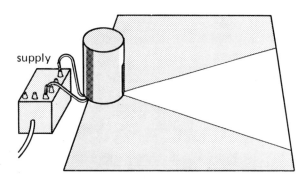

(d) Place the prism in the beam of light as shown below:

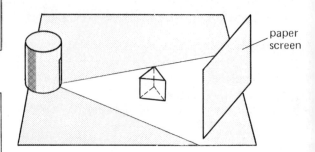

(e) Now hold a piece of white paper roughly in the position shown above. Notice the colours at the edges. If these are faint, hold some black paper over the top, as a shield.

▲ What to write
1. What do you see on the paper screen?
2. Where has the light on the paper screen come from?

★ Information
In fact, white light is made up of a mixture of colours. The prism separates these colours.

▲ What to write
3. From what you have seen, what colours does white light consist of?

Unit 9
Mixing colours

■ **What to do**

(a) Switch on the colour mixer.

(b) Look carefully at the effect produced.

▲ **What to write**

1. What did you see as the colour mixer was spinning?
2. Can you say *why* this happened?

★ **Information**

White light is made up of the colours of the *spectrum*:

red, orange, yellow, green, blue, indigo, violet

Our colour-mixer mixes the main colours together again to give (nearly) white.

▲ **What to write**

3. Where in nature have you seen a spectrum like this? (Hint: Look at the diagram below of a raindrop in the Sun's rays.)

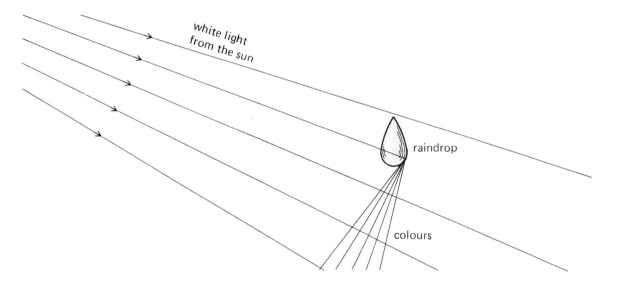

2

And There Was Light...

Unit 1
Better effects with prisms

★ **Information**

It is interesting to look closer at the *spectrum* caused by a prism.

■ **What to do**

(a) Collect:
a light source
a low voltage supply
a card with one thin slit
(as shown below)
a card without a slit
a prism
a large sheet of white paper

(b) Lay the large sheet of white paper flat on the table-top.

(c) Set up the light source and card with the slit so that a single narrow ray of light is projected across the paper on the table:

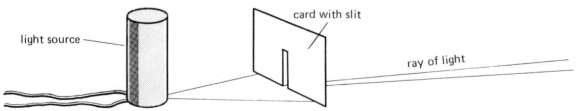

(d) Now place a prism at an angle to the ray, and position the other piece of card as a screen to "catch" the spectrum:

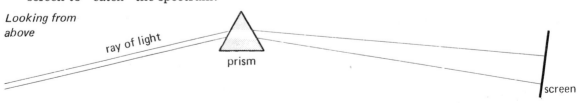

▲ **What to write**

1. What happens to the single narrow ray of light in the prism?
2. What are the main colours that the ray of light is made up of?
3. Which colour is bent *most* by the prism?

■ **What to do**

(e) Now get a card with *many* slits (like the one shown below) and put it in place of the single card slit.

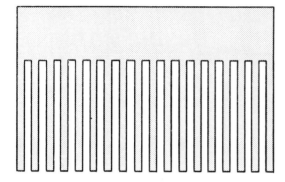

(f) Take the prism away, and place a cylindrical lens . . .

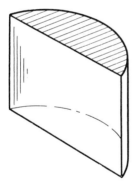

. . . in the rays of light, like this:

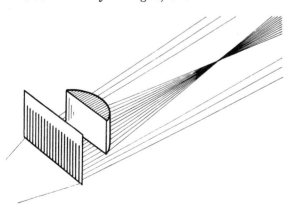

(g) Notice the point where the rays come together. That is the lens' *image point*.

(h) Adjust the position of the lens so that the image point is 15 – 20 cm from the lens.

(i) Now place the prism in the light that is coming from the lens, and hold the card screen, as before, about 15 cm from the prism.

▲ **What to write**

4. Do you get a clearer spectrum this time?
5. Try drawing a *plan view* of the rays being bent by the lens, *before* you put the prism in the way. Copy and complete the diagram below.

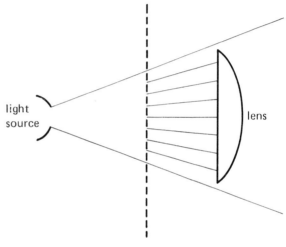

6. Try drawing a *plan view* of the rays being bent by the prism. Copy and complete the diagram below.

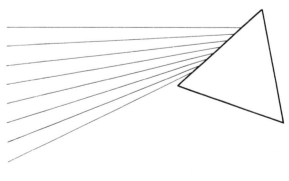

Unit 2
More about images

■ **What to do**

(a) Keep your apparatus from Unit 1 set up, except that you no longer need the prism.

(b) Fetch a *second* cylindrical lens.

(c) Put the second lens in line with the first, as shown below:

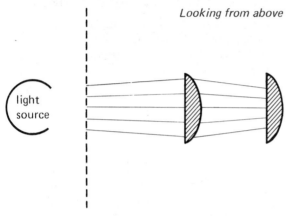

Looking from above

(d) Notice what happens to the *image point*.

▲ **What to write**

1. What happens to the image point when you put in the second lens?

2. Copy and finish off the diagram above, showing what happens to the light going through the second lens.
   (Hint: Make your drawing *large*, and use a *ruler* to draw in the light rays.)

■ **What to do**

(e) Now try two different lenses in turn instead of the first two lenses:

First, try a thicker, more curved lens, which looks like this:

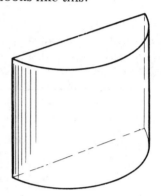

Now, try a strange-looking lens ... like this:

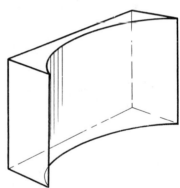

▲ **What to write**

3. Does the thick lens bend the light more than this lens?
   Would you say it is "stronger"?

4. Try to draw what the strange-looking lens does to the light.
   (Draw another plan view.)

Unit 3
How a lens forms an image

★ **Information**

We know so far that lenses that are thicker in the middle than at the edges (called *convex* or *converging* lenses) can form clear, small, upside-down images of distant bright objects on a wall or screen held a certain distance away from the lens.

We can use the square "cylindrical" lenses from Unit 1 of this chapter to show us how these images are formed.

▲ **What to write**

1. What is the special name given to a lens that is thicker in the middle than at the edge?
2. What sort of image does it form of bright, distant objects?

■ **What to do**

(a) Fetch a large sheet of white paper and lay it flat on the table-top.

(b) Fetch a large sheet of black paper
    a light source
    a low voltage power supply
    a card with many slits ("comb")
    a cylindrical lens
    a piece of card.

(c) Set up the light source and card with many slits so that there are many clear rays of light spreading out across the paper:

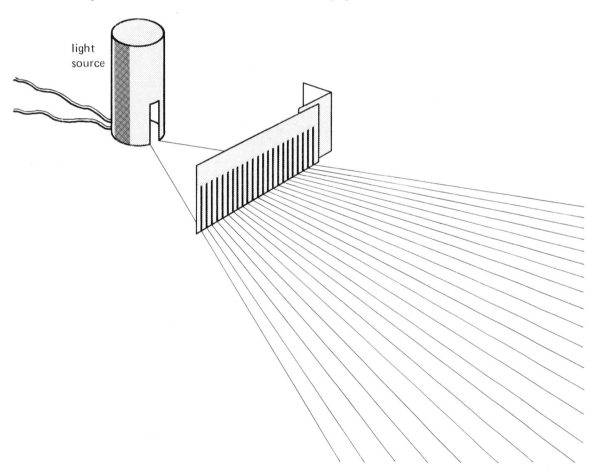

(d) Set up the lens in the light rays, so that it brings the rays together at an image point.

(e) Hold the piece of card just in front of the lens, blocking off all the rays going through it, except one on the edge of the lens:

(f) Now move the card sideways, gradually unblocking each ray in turn, until all the rays are once more going through the lens.

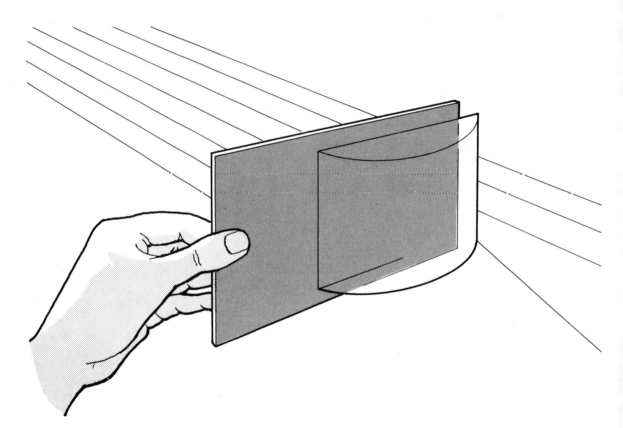

▲ **What to write**

3. From your careful observation, can you say:

   i. which ray is refracted most by the lens?

   ii. which ray is refracted least by the lens?

4. Is there a ray that is not refracted at all as it passes through the lens?

5. Do the rays of light stop at the image point, or do they carry on beyond it?

6. What happens to the distance between the lens and the image point as you move the lens:

   i. closer to the light source?
   ii. further away from the light source?

Keep your apparatus for the next Unit.

Unit 4
Finding out why the image is upside-down

■ What to do

(a) Position the lens on the table so that it produces an image point on the paper roughly 10 cm from it.

(b) Now move the comb right up to the lens, positioning it about 2 - 3 cm from the lens between the lens and the light source.

(c) Fetch a *second* light source and position it next to the first, directing its beam of light along the table towards the lens.

(d) Adjust the positions of the pieces of apparatus until you get *two* clear image points.

(e) Spend some time experimenting with your apparatus, answering the following questions as you go along.

▲ What to write

1. Get a sheet of unruled paper, and draw a plan view of the experiment, filling the whole page, showing the light rays, and drawing all the straight lines *using a ruler*.

2. Look at the diagram below. Which image point is caused by which light-source?

3. As you move light-source A sideways as shown, which image point will move, (X or Y), and in which direction?

4. Try to explain carefully why an image caused by a convex lens is usually upside-down.

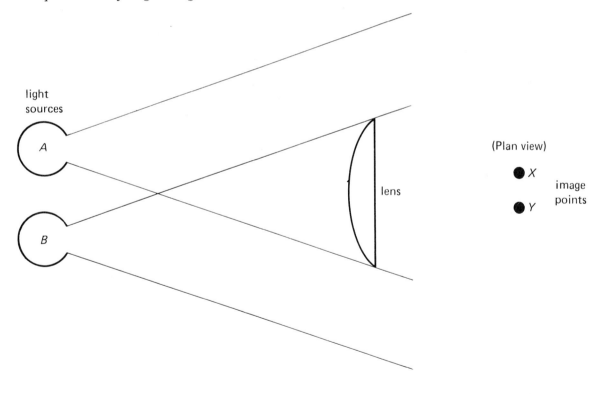

15

Unit 5
A model telescope on the table-top

★ **Information**

You will need a lot of room for this experiment — try to find a long table to do it on.

■ **What to do**

> (a) Collect:
> a cylindrical +7D lens (thin)
> a cylindrical +17D lens (thick one)
> two light sources
> a card with three slits
> a large sheet of white paper
> a low voltage power supply.

(b) Lay the white paper down on the table as before.

(c) Set up the +7D lens with the three slit card about 2–3 cm from it, and the two light sources a long way away down the table, behind the card.

(d) Adjust the positions of lens, card and light sources, so that you can get the two image points about 10 cm from the lens, as in Unit 2.

(e) Now use the +17D lens as a magnifying glass. Place it quite close to the two image points, in line with the other lens and light sources, so that the image points are between the two lenses.

(f) Use your apparatus to answer the following questions.

▲ **What to write**

1. How does the image formed by the +7D lens (represented by the two image points) compare with the bright object (represented by the two light sources)?

2. If you look along the table-top, through the two lenses, do the two light sources appear further apart than they really are?

3. Does the +17D lens make an image point on the paper?

4. Look at the rays coming out of the "eyepiece" lens (+17D). Is there a point from which they *seem* to be coming, if they are traced backwards in straight lines?

★ **Information**

An *image* that we can collect on paper, to form a picture, or can actually be seen where rays of light come together, is only *one* type of image. It is called a *real* image, because we can actually "collect" it on a screen.

Another type, called a *virtual* image, cannot be collected on a screen, or be seen where rays come together, but it does exist. We see a virtual image when we use a lens as a magnifying glass, or as a telescope eyepiece.

If we hold a convex lens very close to an object, and look through it the right-way-up, the magnified image we see is a virtual image.

In the telescope eyepiece, we can trace back the rays coming from the eyepiece and find out the size and position of the virtual image we see when we look through it:

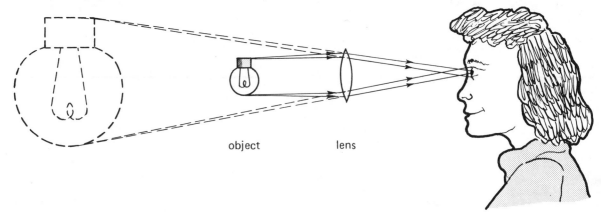

object   lens

Unit 6
Another virtual image

■ **What to do**

(a) Have another look at the —17D lens in the rays of light from a light source. Look for a virtual image by looking through it, and tracing the rays back.

★ **Information**

This sort of lens is called a *concave* or *diverging* lens. It is thicker at the edge than in the middle.

▲ **What to write**

1. Does the —17D lens cause a virtual image?

2. Why do you think we usually call the convex lens a *converging* lens, and the concave lens a *diverging* lens?

   (Hint: "Converging" means roughly "coming together", "diverging" means roughly "spreading apart")

Unit 7
More about reflection

■ **What to do**

(a) Collect:
   - a light source
   - a low voltage supply
   - a multi-slit card (comb)
   - a plane mirror and holder,
   - a large sheet of white paper
   - a large sheet of black paper.

(b) Make a "fan" of rays of light on the table top as in the previous Units.

(c) Set up the mirror in the rays of light, so that you can see the rays that reflect off it. (The mirror should be upright, at about 45° to the rays.)

(d) Hold the black paper over the apparatus so that you get a clear bright light-ray pattern.

(e) Look along the rays of light reflected off the mirror, and look for any virtual images.

▲ **What to write**

1. Whatever you see in a lens or a mirror is really a sort of image. Is the image you can see in the mirror *real* or *virtual*? Why?

2. Where do the rays of light bouncing off the mirror appear to come from, if traced backwards in straight lines through the mirror?

3. Look at the diagram below. It tells you what we call the *angle of reflection* and the *angle of incidence*. Are they always the same as each other? Try turning the mirror to a different angle and see.

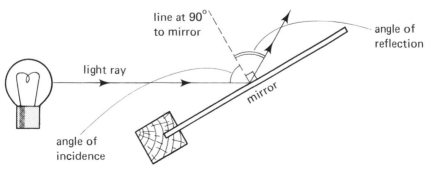

3

Camera and Eye

★ **Information**

We have already thought about some types of *optical instruments*. The most important optical instruments to us are our eyes: we can *see* things that reflect light to our eyes.

To get a good idea of how our eyes work, we can have a look at another type of optical instrument which, in many ways, is similar to the eye.

This is the *camera*

Unit 1
Making a pinhole camera

■ **What to do**

(a) Collect:
a pinhole camera box
one small piece of black paper
one large piece of black paper
a small piece of greaseproof paper
a pin.

(b) Tape a square of greaseproof paper over the back of the box, and the small piece of black paper over the circular hole in the front:

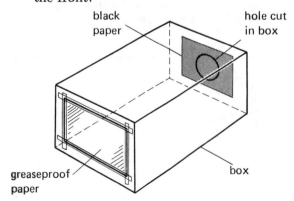

(c) Prick a pin-hole in the centre of the black paper.

(d) Fasten the large sheet of black paper round the screen end to shield the back of the camera from light in the room:

(e) Hold your pin-hole camera with the pin-hole end pointing towards a *bright* object.

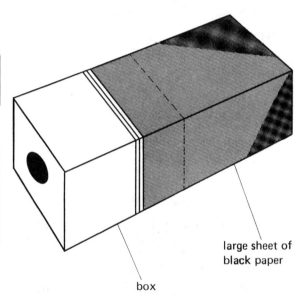

(f) Look at the greaseproof paper screen.

▲ **What to write**

1. What do you see on the greaseproof paper screen?

2. Describe what you see:

   Is it smaller than the bright object?
   Is it very dim?
   Is it clear or sharp?
   Is it upside-down?

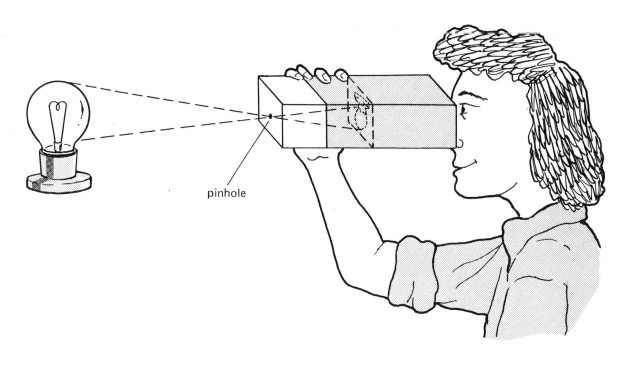

pinhole

★ **Information**

So far, the image you can see on the screen is far too dim. For a real camera, we need a much brighter image. To get it, we must *let more light in*.

The problem is, how do we do it without ruining the picture?

Unit 2
Letting more light in

■ **What to do**

(a) Prick two more pin-holes in the black paper of your camera, like this:

(b) Use your camera as before.

▲ **What to write**

1. What did you see now on the greaseproof screen?

2. Were there as many images as there were holes?

Unit 3
Trying another idea

■ **What to do**

(a) Now make a larger hole in the black paper of about the thickness of a pencil instead of all the tiny pin-holes.

(b) Repeat the test of your camera.

▲ **What to write**

1. Now what do you see on the greaseproof screen?

2. Was the image brighter than with a single tiny pin-hole?

3. Was the image larger than with a single pin-hole?

4. Was the image nice and clear?

5. Can you suggest something, instead of a large hole, that will give *both* a *clear* image and a *bright* image? (You have come across it before.)

Unit 4
Making a lens camera

■ What to do

(a) Get a large spherical lens.

(b) Cut a round hole in the black paper of your pin-hole camera, a little smaller than the lens.

(c) Carefully tape the lens to the front of your box over the hole — make sure it is firm and will not fall off.

(d) Now test the camera as you did in the other Units.

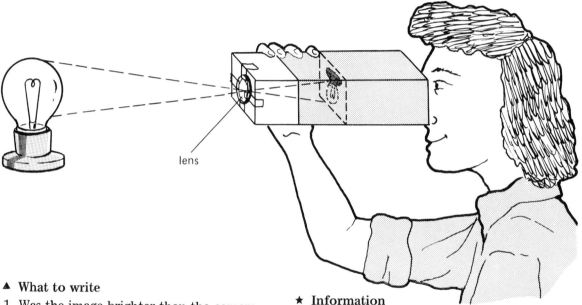
lens

▲ What to write

1. Was the image brighter than the camera with the tiny pinhole? if so, why

2. Was the image clearer than with a large hole?

3. Write about a method by which you could *adjust* the brightness of the image of your lens camera, without changing the lens. Try out your ideas to see if they work.

★ Information

In the later Units on the eye, think about this work on the pin-hole camera, and think about:

 i. things that are the same between the camera and the eye;

 ii. things that are different.

Unit 5
Eye dissection

★ **Information**

You may be allowed to dissect an eye. You must follow the instructions given by your teacher very carefully.

The drawings below will help you see the various parts of the eye.

When you have finished the dissection, answer the questions.

▲ **What to write**

1. Why is it important to wash your hands after the dissection?

2. You will have noticed the lens in the eye. What is the lens for?

3. Can you use the lens as a magnifying glass, like you did with the glass lens?

4. Very curved lenses focus on close things. Thin lenses focus on far away things.

    Why is it important for the eye lens to be able to change its shape?

5. The 'Optic Nerve' runs from the eye to the brain. What do you think it is for?

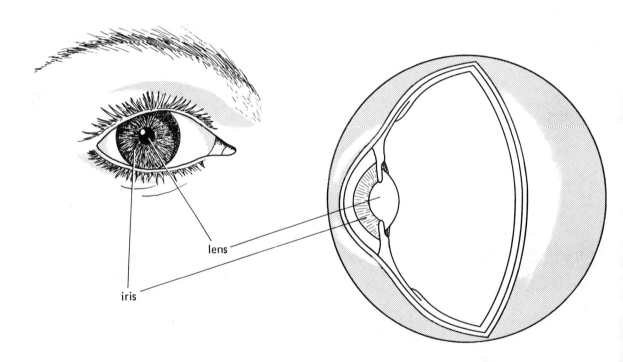

Unit 6
Eye protection

▲ **What to write**

1. Why, do you think, are your eyes set back into your head, surrounded on most sides by bone?

★ **Information**

To help you with questions 4 and 5, think of how a car driver keeps his windscreen clean.

▲ **What to write**

4. You continuously produce tears, to keep the front of your eyes moist. (You don't have to be crying to make tears.) Why are these tears necessary?

5. Why do we blink from time to time?

■ **What to do**

(a) Shine a light into your partner's eye. Look at the size of his pupil.

(b) Switch the light off, and continue to look at the pupil.

(c) Repeat this several times.

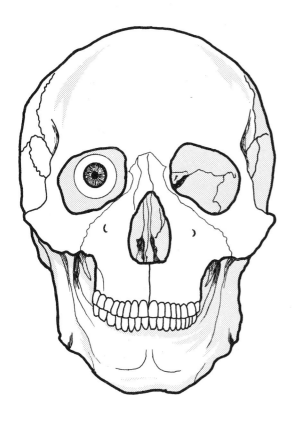

▲ **What to write**

6. What happened to the size of the pupil, when you switched the light off?

7. Can you suggest why our eyes do this? (Remember how you made the picture in the pin-hole camera lighter.)

8. When you go into a cinema, it looks dark, and the usherette has to show you to your seat with a torch.

   But you can easily leave without help. What has happened to the size of the pupils to make this possible?

9. Why is it that car lights, or street lights, look dull in the day but very bright at night?

2. If somebody moves something quickly towards your eyes, (like a fist!) what do you do, without thinking, to protect your eyes?

3. If some dust actually gets onto your eye, your eye begins to water a lot. How does this help?

Unit 7
What are eyes for?

▲ **What to write**

1. You use your eyes for 'seeing' things. But what do your eyes actually tell you *about* the things?

   Do they tell you

   i. An object's size?
   ii. It's colour?
   iii. It's smell?
   iv. Distance?
   v. Noisiness?
   vi. Brightness?
   vii. It's 'feel'?
   viii. Temperature?
   ix. It's shape?

   Write down what they tell you.

■ **What to do**

(a) Stand on tip toe, on *one* leg, and try to keep your balance for 1 minute.

(b) After a rest, repeat the exercise, only this time, *keep both eyes firmly shut.*

(c) Now try it again, with your eyes open.

▲ **What to write**

2. Was it easier to keep your balance with your eyes open, or with them shut?

3. What did other people think?

4. Can you explain how, by keeping your eyes open, you can keep your balance better?

■ **What to do**

(d) Copy the results table shown below.

	No. of times you hit the cross
Both eyes open	
Right eye shut	
Left eye shut	

(e) Mark a cross on a piece of rough paper.

(f) Place the paper 2 – 3 feet away.

(g) Look at the cross carefully, and then, fairly quickly, try to put your index finger down on the cross.

(h) Move the paper a little, and then repeat the exercise 5 times. Record in your table the number of times you hit the cross.

(i) Now repeat everything, but with your right eye shut

(j) Repeat it again, but this time with your left eye shut.

▲ **What to write**

5. Was it easier to hit the cross with both eyes open, or with one eye shut?

6. When was it most difficult, with your right eye shut, or with your left eye shut?

7. Does it seem to be true that we judge distance better with both eyes open?

Extension Work 1
Illusions

■ **What to do**

(a) Look at the diagrams below. In each case try to see which line, *a* or *b*, is the bigger.

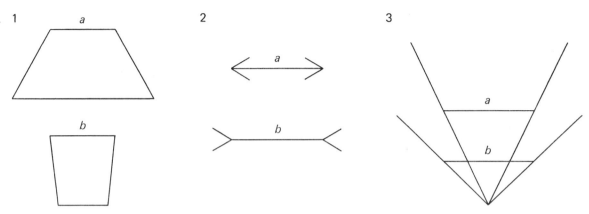

(b) Measure the lines, using a ruler.

(c) Look at the diagram of the fence, below. Estimate which fence post is the biggest. Also measure them.

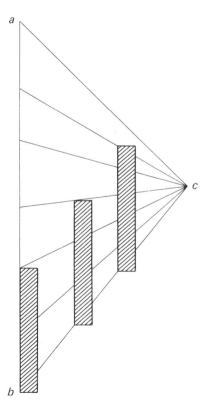

(d) In the diagrams below, are the horizontal lines parallel?

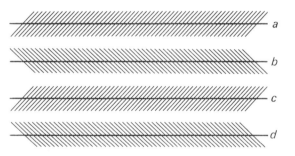

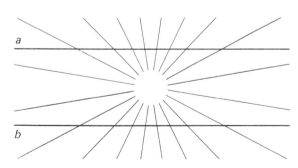

(e) You have seen six examples of 'optical illusions'. Do you know any more, or can you invent any?

Try them out on your teacher or your friends.

Extension Work 2
Eye positions

■ **What to do**

(a) Look at the pictures of animals below.

▲ **What to write**

1. Write down the names of the herbivores (plant eaters).

2. Write down the names of the carnivores (flesh eaters).

3. What do you notice about the position of the eyes of herbivores, compared with carnivores?

4. Can you explain why herbivores usually have their eyes on the side of their heads?

5. In your earlier work, you found that we are better at judging distance if we can see an object with both eyes. (Look back at Unit 7.)

   Suggest why the carnivores have their eyes at the front of their heads.

■ **What to do**
(b) The same applies to birds. Look at the position of the eyes in the birds of prey. Notice the large area of binocular vision.

4

Forces

Unit 1
What is a force?

★ **Information**

In everyday use, the word 'force' can mean a lot of different things

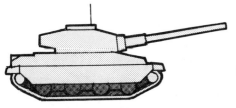

In science, we have a very definite meaning for the word 'force'. We just mean any 'push' or 'pull', nothing else.

Scientists show the direction of a force with an arrow:

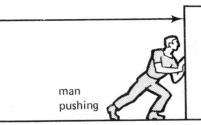
man pushing

We draw a *long* arrow for a *large* force, and a *short* arrow for a *small* force.

small arrow — small force

large arrow — large force

▲ **What to write**
1. Copy the drawings below and draw arrows to show the pulling force acting:

   i. on the donkey

   ii. on the fish

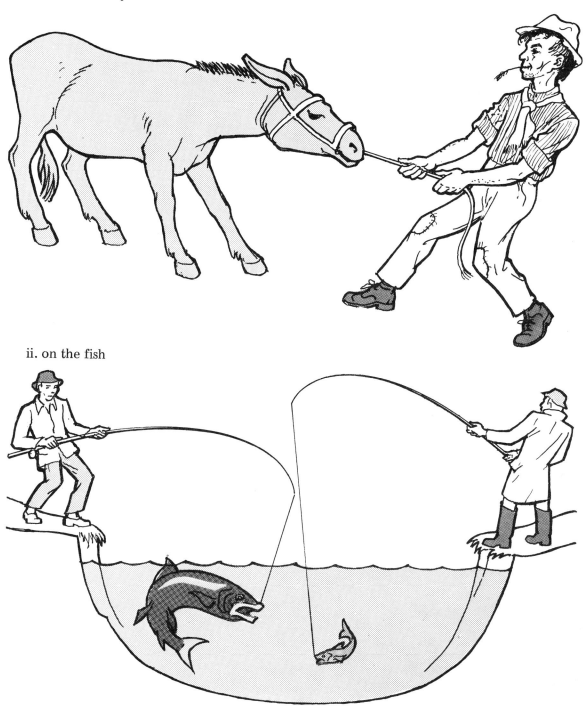

2. For which fish did you draw the longest arrow?

Unit 2
Forces exhibition

★ **Information**

Look at the exhibits in any order and have your work marked after each one.

★ EXHIBIT 1   FORCES ON MAGNETS

■ **What to do**

(a) Hold the *north pole* of the floating magnet (the red end) next to the *north pole* of the fixed magnet (marked N).

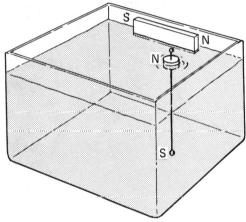

(b) Release the floating magnet and watch the path that it takes.

(c) Repeat this several times.

▲ **What to write**

1. The diagram below shows the magnets in the water, looking from above. Copy the drawing and show the paths taken by your floating magnet when you released it.

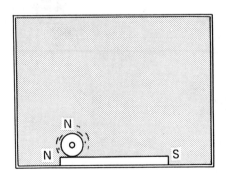

★ **Information**

Instead of 'pushing' and 'pulling', in work with magnets we generally use the words *attract* and *repel*.

e.g. when two magnets try to push each other away, we say they 'repel each other'.

▲ **What to write**

2. Choose the correct word (attracts or repels) to fit into the following sentences and copy the sentences into your book.

   i.   A north pole .... a north pole
   ii. A south pole .... a south pole
   iii. A north pole .... a south pole
   iv. A south pole .... a north pole

★ EXHIBIT 2   ELECTRICAL FORCES

★ **Information**

Polythene that has been rubbed with a duster has a negative electrical charge on it.

Cellulose acetate when rubbed has a positive charge.

(Instead of 'pushing' and 'pulling', in work with electrical charges we use the words *repel* and *attract*.)

■ **What to do**

(a) Give the suspended, white polythene strip some negative electrical charge by rubbing it with a duster.

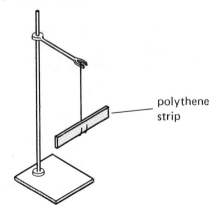

polythene strip

(b) In the same way give another strip some negative charge.

(c) Hold the second near the first as shown, and notice whether they attract or repel each other.

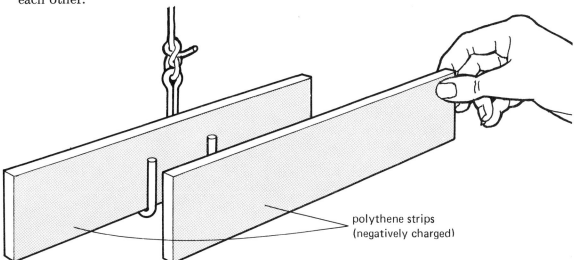

(d) Repeat the experiment using a cellulose acetate strip which has been given some positive charge by rubbing.

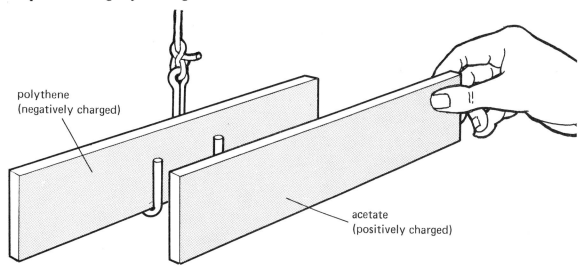

▲ What to write

1. Do the negatively charged polythene strips attract or repel each other?

2. What happens when a positive charge is brought near a negative charge?

3. 'Like charges repel, unlike charges attract'. Use your results to explain what this means.

■ What to do next

(e) Charge up a polythene strip by rubbing.

(f) Hold the strip near to a small jet of water from a tap.

▲ What to write

4. Use a drawing to show what happens to the jet of water.

★ EXHIBIT 3 FRICTION FORCES

★ **Information**

Friction forces are forces which try to stop a thing moving

■ **What to do**

(a) Drop a nail into the jar containing glycerol and watch it sink.

▲ **What to write**

1. Can you suggest how the particles in a liquid try to stop something from passing through it?

2. Is it easier to run against the wind or with the wind? Why is this?
   (Hint: think of the speed of the air particles hitting you.)

3. Why are aeroplanes streamlined, but not spacecraft which orbit the Moon?

■ **What to do**

(b) Note the pull required, for both the smooth surface and the rough, when a block is pulled at a steady speed from one to the other.

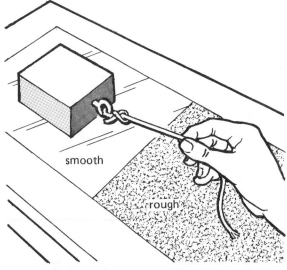

(c) With a hand lens take a look at
   i. the rough surface
   ii. the smooth surface
   iii. the surface of the block.

▲ **What to write**

4. Which diagram below shows the direction of the friction force?

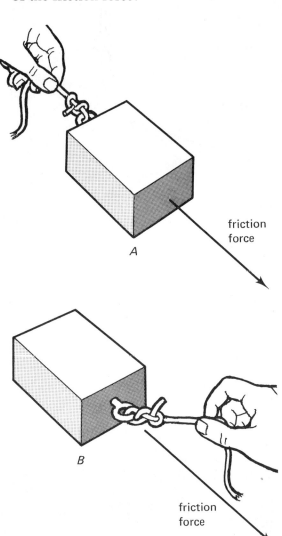

5. Is the force greater or smaller when pulling the block over a rough surface?

6. Were the surfaces of the block and the smooth paper really smooth?

7. Do a drawing to show why surfaces have difficulty in sliding over each other.

8. In car engines, what do we use to reduce friction between the moving parts?

★ EXHIBIT 4   SPINNING FORCES

■ What to do

(a) Hold a heavy mass in each hand and sit on the turntable.

(b) Stretch out your arms and give yourself a spin.

(c) While spinning, bring your hands close in — notice what happens.

(d) Stretch out your arms again — notice what happens.

▲ What to write

1. What happens to the spinning speed when you bring in your arms?

2. What must your arms do to stop the masses shooting away, pull them in or push them out?

★ EXHIBIT 5   ELASTIC FORCES

■ What to do

(a) Hold the luggage strap in each hand and stretch it — notice the pull in each hand.

(b) Keep your left hand fixed and increase the stretch with your right.

▲ What to write

1. Is the pull on each hand the same or is it different? If you are not sure try the following:

Forcemeters

2. When you have increased the stretch is

   i. the pull the same in each hand?

   ii. the pull greater or smaller than before?

★ EXHIBIT 6 GRAVITY FORCE

★ Information

Every object on Earth feels a downward gravity force on it due to the Earth's pull.

The gravity force on a mass of 1 kg is about 10 newtons.

The gravity force pulling on an object is usually called weight.

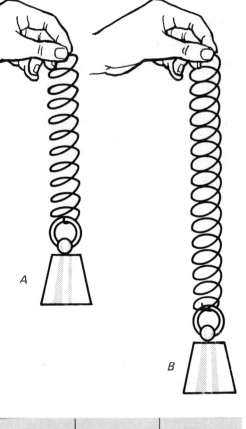

■ What to do

(a) Using the *same* spring, first suspend mass A and then mass B.

▲ What to write

1. Just by looking at the stretch of the spring, how would you have known which mass had the largest gravity force pulling it?

2. What word means the same as 'gravity force'?

3. Draw a result table like the one shown —

Mass (grams)	200	400	600	800	1000 (1 kg)
Gravity force (weight) in newtons					

■ What to do

(b) With a forcemeter, measure the gravity force in newtons on the masses written in your table.

(c) Measure the gravity force (weight) on the masses A and B.

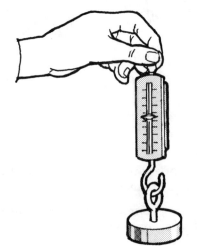

▲ What to write

4. What are the masses of A and B? (Use your results table to help you.)

Unit 3
Balanced and unbalanced forces

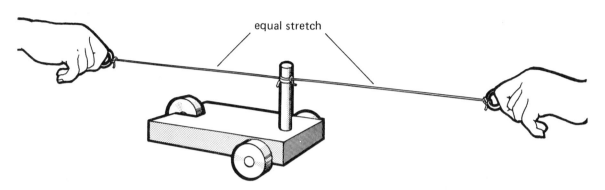

■ What to do

(a) Obtain two elastic cords and a dynamics trolley.

(b) Loop the ends of the cords over the post on the trolley.

(c) Stretch both cords by *equal* amounts.

▲ What to write

1. Are the forces pulling on the trolley equal? Explain your answer.

2. Does the trolley move?

3. Do a drawing of the trolley and include arrows to show the forces acting on it.

4. Copy the following sentence choosing the right words.

   If an object is stationary, all the forces acting on it *balance/do not balance*.

■ What to do

(d) Reduce the stretch on one cord, keeping the stretch on the other cord the same as before.

   When the trolley moves, try to keep the new lengths of stretched cord from changing by moving your hands along with it.

▲ What to write

5. To make the trolley move, are the lengths of the stretched cords the same or different?

6. If the trolley moves, are the forces balanced or unbalanced?

7. Copy the following sentence, choosing the right words.

   The unbalanced forces acting on the trolley made it *stay where it was/speed up/slow down*.

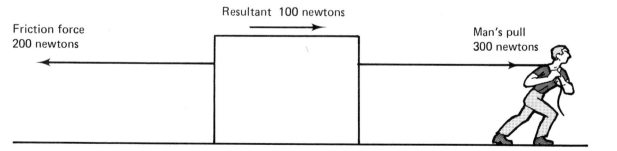

35

★ **Information**

When there are two or more forces pushing or pulling on an object *and they do not balance*, we say there is a *resultant* or overall force acting on it.

▲ **What to write**

8. Look at the drawings below of a tug-of-war, and for each one decide

   i. the size of the resultant force
   ii. which way it points (to the left or right)
   iii. which way the flag is moving.

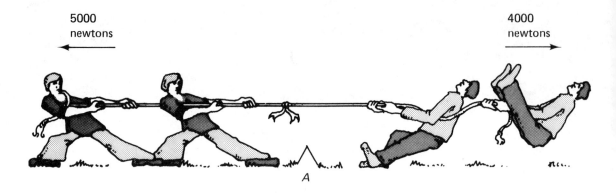

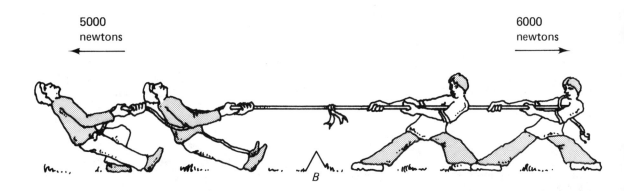

Unit 4
Why is it difficult to get things moving?

★ **Information**

As well as forces such as gravity, friction etc, there is something else which makes it difficult to start something moving. We call it *inertia*.

Inertia is not a force. It is a sort of 'I-want-to-stay-where-I-am-ness' caused by the object's size or mass.

■ **What to do**

(a) Collect:
  a postcard
  a metal washer
  a plastic container
  2 wooden blocks (one large, one small)
  2 elastic cords

(b) Lay the postcard on the beaker and put the washer on top.

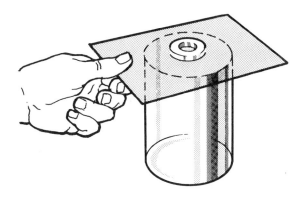

(c) Sharply pull the card away

▲ **What to write**

1. What happens to the washer?
2. Why doesn't the washer move along with the card?

■ **What to do**

(d) Set up the apparatus below.

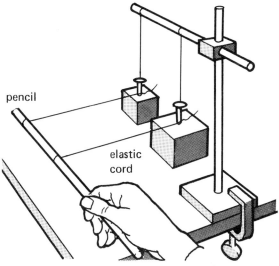

(e) Give the pencil a sharp jerk. This should not be too violent -- just sharp enough to make *both* blocks move.

(f) Note the different amounts of stretch needed to get each block moving.

▲ **What to write**

3. Which block moved first?
4. Which block had the largest inertia?
5. Which block needed the greatest force to get it moving?

Unit 5
What do we mean by pressure?

★ **Information**

A lot of people think the word 'pressure' has the same meaning as the word 'force'.

To scientists there is a difference, as we shall see.

■ **What to do**

(a) Collect a clamp and a large wooden block with a nail in.

(b) Hold the apparatus as shown.

(c) Place your hand underneath so that the nail is resting on your palm — remember what this feels like.

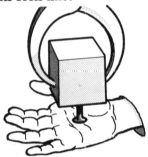

(d) Turn the block upside down in the clamp and notice how different it feels now.

(e) Instead of using your hand, repeat the experiment with a block of Plasticine.

▲ **What to write**

1. Which side presses more into your hand, the side with the nail or the side without it?

2. Was the gravity force (the weight) of the block and clamp the same for both positions? (Test with a forcemeter if you are not sure.)

★ **Information**

In the experiment that you have just done you should have found that, although the weight did not change, the nail side of the block pressed more into your hand.

This is because the gravity force was concentrated into the small area of the nail head, whereas the other side spreads this force out more.

So the pressure was greater under the head of the nail.

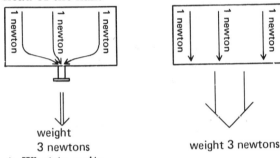

weight 3 newtons   weight 3 newtons

▲ **What to write**

3. Below are examples where we deliberately try to increase or decrease the pressure. For each case say why we do this.

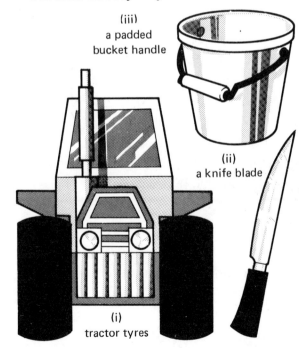

(iii) a padded bucket handle

(ii) a knife blade

(i) tractor tyres

Extension Work 1
The model suspension bridge

Making the bridge

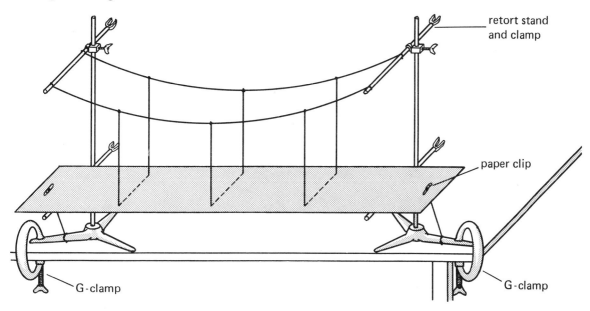

- **What to do**

(a) Collect:
   a sheet of cardboard
   8 short strings
   2 long strings
   2 stands and 4 clamps
   2 G-clamps
   a metre rule
   a nail.

(b) With the nail, make small holes in the piece of cardboard.

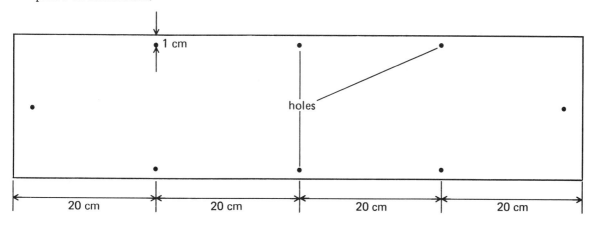

(c) Screw the clamps to the stands, and with the G-clamps, clamp the stands to the bench.

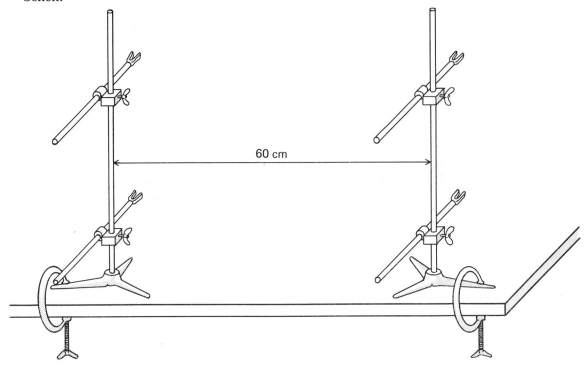

(d) Place a metre rule over the bottom clamps and lay the card on top.

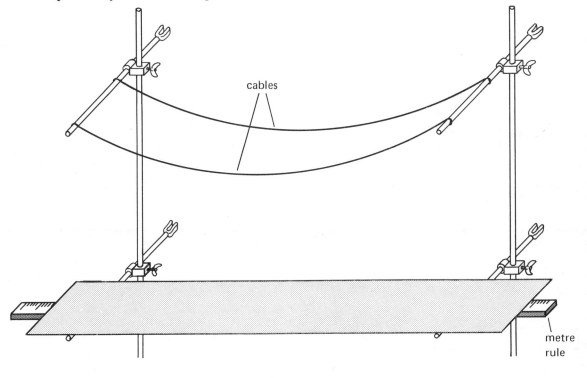

(e) Tie two pieces of string (the cables) to the upper clamps — these must sag in the middle.

(f) Tie vertical strings to the cables as shown below:

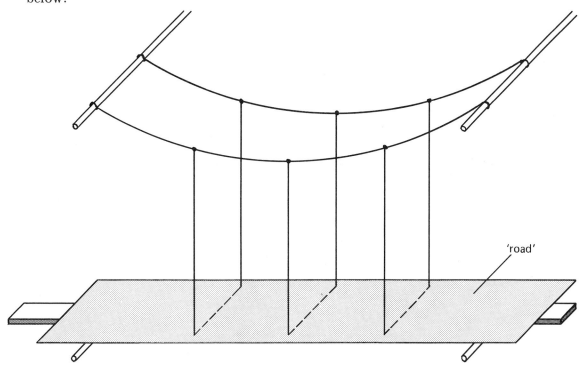

(g) Remove the ruler and check that the road is level.

(h) Tie a piece of string to the base of each stand and pass it through the holes at the ends of the road — tie paper clips to the ends.

Extension Work 2
Testing the bridge

■ **What to do**

(a) With a forcemeter, measure the force in the cables when a 1 kg mass is placed first at one end and then the other.

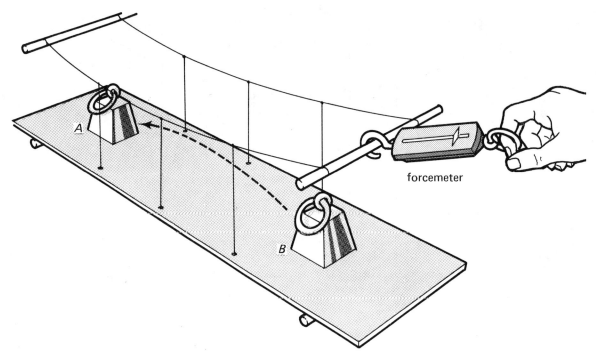

(b) Find out the largest load that the bridge will carry by spacing out 100 g masses along the road.

▲ **What to write**

1. What was the largest load that the bridge would carry?

2. What was the force in the cables at one end when the load was placed:

   i. near that end;

   ii. at the other end?

3. If the bridge were overloaded at the end A (see diagram above), which end of the cables would you expect to break first?

4. Give the names of one or two bridges which are built like this.

Extension Work 3
How to work out pressure

★ **Information**

To work out the pressure on a surface we can use this equation —

Pressure = Force in newtons ÷ Area

Example

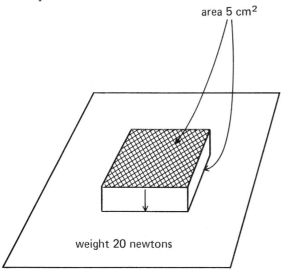

$$\text{Pressure} = \frac{20 \text{ newtons}}{5 \text{ cm}^2} = 4 \text{ newtons/cm}^2$$

(4 newtons per square centimetre)

■ **What to do**

1. Work out the pressure on the ground in newtons/cm²

   i. for a man wearing snow shoes

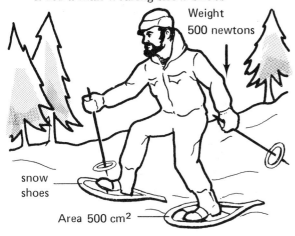

   ii. for a man wearing boots

   iii. on the ice for a woman wearing ice skates

2. Is there a greater or smaller pressure under a snow shoe than under a boot? What advantages do snow shoes have?

★ **Information**

A very high pressure makes ice melt, and the water which is formed makes the surface slippery.

▲ **What to write**

3. Explain why an ice skate slides on ice.

To the Moon

Unit 1
Getting away from it all

★ **Information**

5 - 4 - 3 - 2 - 1 - - - 'BLAST OFF' . . . Imagine that you are on board the Apollo 11 spaceship which took off on July 16, 1969, to land the first man on the Moon. Your companions are Neil Armstrong, Edwin Aldrin and Michael Collins.

The gigantic Saturn 5 rocket is used to propel the spacecraft away from the Earth. It is divided up into sections or stages, each of which 'fires' for a time and then drops away, until the spacecraft has escaped most of the Earth's gravity pull. The spacecraft then travels the rest of the way by itself.

In this section of work, we are going to think of some of the problems of getting to the Moon, and what it is like when we get there. If you use your imagination, you should find that it helps with a lot of difficult ideas.

▲ **What to write**

1. What is the pull of the Earth's gravity on the Apollo spacecraft (46 000 kg) in newtons?
2. We have to lift off from the Earth on top of the enormous Saturn 5 rocket. Why don't we use quite a small rocket?

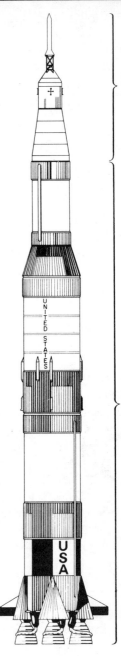

Apollo space-craft mass 46 000 kg

Saturn 5 rocket — uses liquid oxygen, liquid hydrogen and paraffin as fuel.

Unit 2
Out in space

★ **Information**

We have now left the Earth and we are on our way to the Moon. Because we are a long way from the Earth, we can hardly feel the Earth's gravity pull and we are 'weightless'.

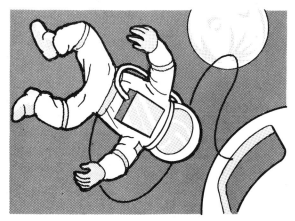

▲ **What to write**

1. Why would an astronaut just float about in 'mid-air' in his spacecraft?
2. Why would he have difficulty in drinking a cup of tea?

★ **Information**

Our spacecraft is in two halves, at the moment joined together — the command module and the landing module.

At first they go round the Moon together. This is called 'being in orbit' round the Moon.

▲ **What to write**

3. In the spinning forces exhibit that you looked at earlier, you should have found that your arms pulled inward on the 1 kg blocks to keep them 'in orbit'. What force keeps the spacecraft in orbit round the Moon?

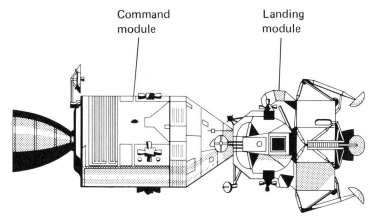

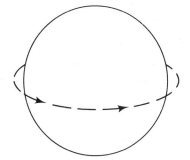

45

Unit 3
The Moon's surface

Imagine that we get into the landing module and go down to the Moon's surface . . .

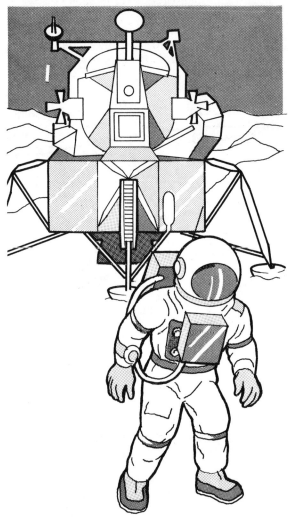

■ What to do

(a) Collect one of the trays marked "Model of a 1 kg block on the Moon's surface".

(b) Collect a *real* 1 kg block.

(c) Imagine that the model in the tray is a 1 kg block actually on the Moon's surface.

(d) Lift up the real 1 kg block and notice its weight. Do the same for the model.

(e) Compare the two weights by lifting them together.

(f) Repeat the experiment, lifting the two blocks with a forcemeter to measure their weights (in newtons).

▲ What to write

1. Copy out and complete these sentences:

   The 1 kg block on Earth weighed . . . . newtons.

   The 1 kg block on the Moon will weigh . . . newtons.

2. Roughly, how many times heavier was one than the other?

3. What force gives things weight?

4. Is the Moon's gravity stronger or weaker than the Earth's?

5. If we had really taken a 1 kg block to the Moon, what would have changed on the way? Would it be --

   i. the size

   ii. the hardness

   iii. the shape

   iv. the amount of metal

   v. the weight

Unit 4
Some facts

If we do take an object to the Moon, the 'amount of stuff' that it is made of will not change (unless we chip a bit off).

The amount of stuff is called the *mass* of the object.

So, mass is measured in kg . . .

It doesn't change!

The force of gravity (the weight) on an object is measured in newtons . . .

It can change!

▲ What to write

1. Say which of these is true or false and say why you think so:

    i. an object can have mass but no weight

    ii. an object can have weight but no mass

    iii. the Earth has no mass

    iv. a 1 kg mass when taken to the Moon becomes about 1/6 of a 1 kg mass.

    v. a 1 kg mass when taken to the Moon weighs 1/6 of what it does on Earth.

Extension Work
Moon questions

▲ What to write

1. The Moon's pull (gravity) is only 1/6th of the Earth's. Can you work out what your weight would be on the Moon?

2. What would happen if an astronaut tore a hole in his space suit?
   (Hint: how would his air supply spread out into a vacuum?)

3. Describe fully what weight is, what it is caused by, and why a 1 kg mass has more weight than a paper clip.

4. The Earth pulls on the Moon — this makes the Moon orbit the Earth. Does the Moon pull on the Earth also? Have you any evidence?

Making Things Work

Unit 1
Motors and engines

★ **Information**

We use motors and engines for a lot of different jobs, and they all use their own special types of fuel.

Sometimes they use fuel to get us about . . .

At other times, motors and engines use fuel to *do a job* rather than to move something about . . .

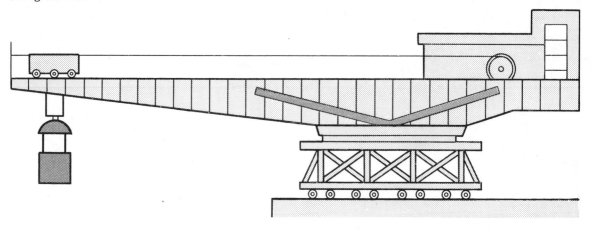

■ **What to do**

Copy Table 1 and write down the type of device (motor, engine etc.) that you would use to do the jobs mentioned.

Table 1

If you wanted to —	What machine or device would you use?	What fuel does it use?
go to London		
go on a Mediterranean cruise		
get to the top floor of a tall building		
deliver newspapers quickly		
play your records		
four more of your own choice —		

**Unit 2
The Human Engine**

★ **Information**

The engines we use every day are our own bodies. Like other engines they use up fuel.

Human fuel = Food

■ **What to do**

(a) Look at Table 2 showing how much food energy we use up for different activities. We measure food energy in kilo-joules, written kJ.

Table 2

Activity	Food energy used (kilo-joules per hour)
sitting quietly	200
brisk walk	800
heavy manual work	2000

(b) Work out the food energy used by a manual worker. This is what he does —

3 hours mixing concrete
½ hour break, sitting quietly
1 hour mixing concrete
1 hour brisk walk home
1½ hours sitting in front of the TV . . .
. . . and then he has his evening meal.

▲ **What to write**

1. How much energy does he use?

2. One potato gives about 920 kilo-joules of food energy. How many potatoes would he have to eat to build up his energy store again?

3. Why would eating potatoes and nothing else be bad for him?

49

Unit 3
How much fuel?

★ **Information**

One of the problems of space flight, and also getting about on Earth, is to find out how much fuel is going to be needed to do what we want to do, and go where we want to go.

For some engines, the amount of fuel energy used up is approximately equal to the amount of work that it does.

For the human body

food energy used = work done

■ **What to do**

(a) Draw Table 3

Table 3

weight of 1 kg mass	newtons
weight of wooden block	newtons
distance lifted each time	metres
time to lift 1 kg mass 5 times	seconds
time to lift wooden block 5 times	seconds

(b) Collect:
  a 1 kg mass
  a wooden block
  a G-clamp
  a forcemeter
  a long piece of string
  a metre rule

(c) With a forcemeter, measure the weights of the 1 kg mass and the wooden block — record them in your table.

(d) Screw the G-clamp to a bench in the lab.

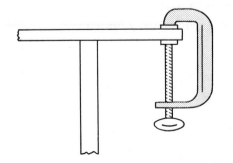

(e) Pass the string over the G-clamp and tie one end to the 1 kg mass which should be resting on the floor or a table.

(f) Pull on the free end of your string, lifting the mass up to the clamp, then lower it to the floor or table again.

(g) With a stop clock, find the time it takes to do 5 lifts (make sure you lift the mass the same distance each time).

(h) Measure the distance you lifted the mass each time and record this in your table.

(i) Repeat the whole experiment using the wooden block.

▲ **What to write**

1. Which made you the most tired, lifting the 1 kg mass 5 times or lifting the wooden block 5 times?

2. In which case do you think you used the most 'human fuel'?

3. You could have tried the experiment again, lifting the 1 kg mass much further.

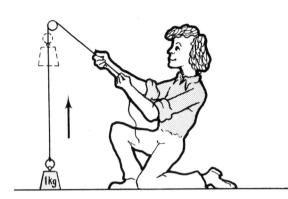

In which case above would you have used the most 'human fuel'?

4. If you tried the experiment yet again, this time lifting a *10 kg mass* the same height as the 1 kg mass, in which case would you have used the most 'human fuel'? In which case would you have done the most work?

5. What does the amount of work we do depend on? Is it . . .
   i. just the force we use?
   ii. the force we use and how far we pull with it?
   iii. the mass of the thing we are pulling?
   iv. whether we can lift it?

6. Find out how much work you did in lifting the 1 kg mass *once* by using your results like this —
   i. work done = force used in newtons (weight) × height lifted in metres
   ii. multiply your answer by five to give the amount of work you did for 5 lifts

★ Information

Your answer to Question 6 will be in newton-metres, written N.m.

A Mini car can do about 20 000 N.m every second. A horse can do about 750 N.m every second.

▲ What to write

7. How much work can you do in every second?
(Hint: divide the amount of work done for five lifts by the time taken for five lifts.)

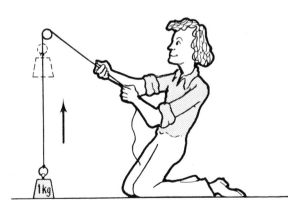

> **Extension Work**
> The electric motor — another worker

■ What to do

(a) Collect:
a small electric motor
a low voltage supply
two leads
some string or twine
some slotted masses
a G-clamp

(b) Clamp the electric motor to the bench with its spindle overhanging the edge.

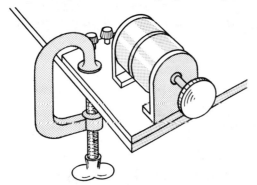

(c) Connect the motor to the red and black terminals of the low voltage supply — make sure red terminals connect to red terminals.

(d) *Have your wiring checked by a teacher before you switch on.*

(e) Switch on and check that the motor is working — switch off.

(f) Tie a length of string or twine to the motor spindle, the other end to a 200 g slotted mass on the floor.

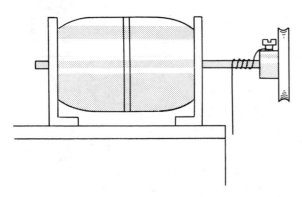

(g) Switch on, making the motor lift the mass.

(h) Try the same experiment using a mass of 100 g.

▲ What to write

1. Which mass did the motor lift easiest?

2. Did it do the same amount of work in each case?

3. What form of energy goes into the motor?

4. Does the motor change this into some other form of energy? What other form?

5. Work out and write down how much work the motor did when lifting the 100 g mass and the 200 g mass.

(Necessary information: 1 kg = 1000 g which weighs 10 newtons)

You will have to measure how far the motor lifted the masses and work out the weight of the masses in newtons.

6. With which mass did the motor use the most electrical energy?

★ Information

From this we can say that 'work' is the amount of energy changed from one form to another.

▲ What to write

7. In view of the information above, do you think that an electric light bulb is working when it is on? If so, describe the energy changes going on.

7

Movement

Unit 1
What gets things going?

■ What to do

(a) Collect:
   a dynamics trolley
   a runway
   a stop-clock
   a metre rule

(b) Make sure the runway is sloping.

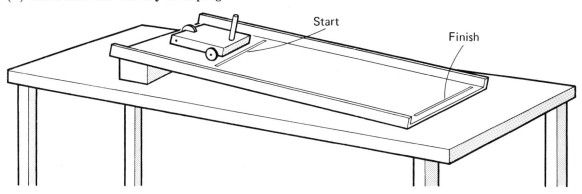

(c) Place the dynamics trolley at the top of this runway, and mark the position of the front wheels (its start position) with a piece of Sellotape.

(d) Put another piece of Sellotape near the end of the runway (the finishing point).

(e) Let go of the trolley so that it runs down the runway. Please do not let it fall off the end. It is easily broken.

(f) Measure how far the trolley runs between the start and finish.

(g) Let the trolley run down the runway again, now timing it over the measured distance.

▲ What to write

1. Write out your results like this —

Distance covered by trolley	metres
Time for trolley to run down the runway	seconds

2. What was the force that made the trolley start moving down the table?

3. Did the trolley move with a steady speed, or did it speed up or slow down?

★ **Information**

When an object speeds up we say it *accelerates*. When an object slows down we say it *decelerates*. If the speed does not change, we say its speed is *constant*.

If it covers a distance in a certain time, we can work out its speed by dividing *distance* by *time taken*.

Example

i. A car goes 60 kilometres in 2 hours, so its average speed is 30 kilometres per hour.

ii. A trolley goes 4 metres in 2 seconds, so it goes at an average speed of 4 ÷ 2 = 2 metres per second.

If an object is not moving with a steady speed, dividing distance by time will give us its 'average' speed.

Average speed = distance ÷ time

▲ **What to write**

4. Did the trolley accelerate as it moved down the table?

5. What was its average speed in metres per second (m/s)?

**Unit 2
Falling objects**

■ **What to do**

(a) Collect a polystyrene ball and a metal or wooden block.

(b) Hold them both at exactly the same height from the ground.

(c) Let them go at the same time and find out which hits the ground first.

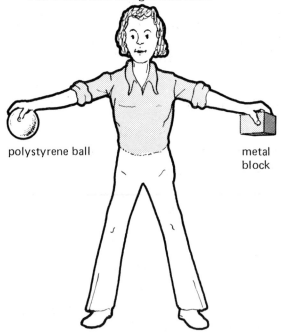

polystyrene ball   metal block

▲ **What to write**

1. Did the objects speed up more quickly (have a larger acceleration) than the trolley on the table? Try to explain why you think this is so.

2. Which is the heavier object, the polystyrene ball or the block?

3. From Question 2, can you say which has the larger pull of gravity on it?

4. Which hits the ground first (or did they both hit the ground at roughly the same time)?

5. Which of these explains Question 4 best:

   i. the ball hit the ground first because it was lighter

   ii. the block had the larger pull of gravity on it, but their different pulls gave them the same acceleration.

   iii. the block hit the ground first because it had less pull of gravity on it and therefore more acceleration.

Unit 3
The accurate measurement of motion

- **What to do**

> (a) Collect:
> a ticker-tape vibrator
> 2 leads
> a power pack
> some ticker tape

(b) Set up a runway so that the trolley accelerates as it goes down.

(c) Place the vibrator at the top of the table, connect it to the 12 V a.c. terminals of the low voltage supply.

(d) Switch on the power pack to check that the vibrator is working. Switch off.

(e) Cut off a length of ticker tape about the same length as the runway and thread one end through the vibrator, underneath the disc of carbon paper.

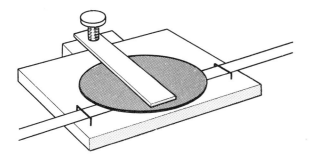

(f) Lick the underneath of the tape nearest the vibrator and stick it to the back of the trolley whilst holding the trolley at the top of the runway.

(g) Switch on the vibrator and let the trolley accelerate down the runway.

(h) Have a close careful look at the pattern of dots drawn on the ticker-tape.

★ **Information**

The vibrator marks off 50 dots per second. You should have a pattern of dots looking like this, only much longer —

Start   Finish

55

▲ **What to write**

1. When are the dots very close together, when the tape is moving fast or slowly through the vibrator?

2. At what point was the trolley moving with its highest speed?

3. What happens to the distance between the dots as the trolley accelerates?

■ **What to do**

(i) If your tape is not very clear, make another tape in the same way as the first.

(j) Look at the fast end of the tape where the dots are spaced wide apart. Count ten spaces between the last dots, and mark this with a pencil line.

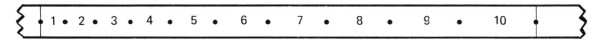

(k) Write the word 'fast' on this section of tape.

▲ **What to write**

4. If the vibrator marks out 50 dots per second, how long does it take to mark out this 10 dot length?

5. What time interval does your ten-tick length represent then? (from Question 1)

6. What length of tape (in centimetres) was pulled through the vibrator during this short time interval?
(Hint: measure the ten-tick length.)

7. How far did the trolley go during this interval?

8. Copy out and complete Table 4

Table 4

Distance travelled per ten-tick time interval	cm
Speed of trolley	cm per ten-tick length

■ **What to do**

(l) Now divide all the tape into ten-tick lengths

(m) Get a large sheet of paper

(n) Draw two lines as shown below — call them the 'horizontal axis' and 'vertical axis'.

```
vertical axis



                              horizontal axis
```

(o) Cut off the first ten-tick length from the slow end and stick it vertically in the bottom left-hand corner.

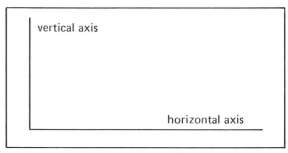

(p) Cut out the next ten-tick length and stick it next to the first. Continue doing this until you end up with a graph looking like this . . .

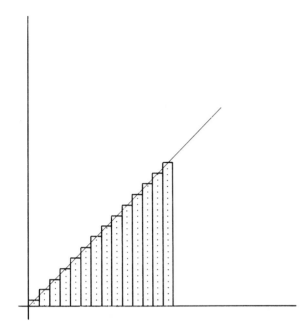

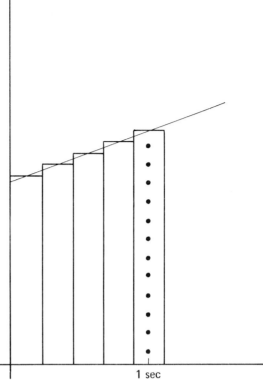

1 sec

▲ What to write

9. What time interval does each ten-tick length represent?

10. We shall make the horizontal axis show time. Each ten-tick tape represents 1/5 of a second. Mark on your time axis the points which represent 1 second, ½ second and 1/5 second.
(Hint: 1 second should be 1 second after the start, so it should be marked at the middle of the 5th tape from the left)
Also show 1½ seconds and 2 seconds.

11. Find on your graph the original 'fast' ten-tick length that you worked out the speed for. Now, *level* with the top of it, mark on the vertical axis this speed. The vertical axis is now showing speed in cm per ten-ticks.

12. Mark on your speed axis points showing the speed of other tapes.

13. What thing about the graph tells you that the trolley is accelerating?

14. Make a quick sketch of what the graph would look like if the speed were steady.

Extension Work 1
Acceleration graphs

▲ **What to write**

1. If you give a dynamics trolley a small push on a level runway, it will speed up as you are pushing it, then, as soon as you let go, it will gradually slow down and eventually stop.

   i. Is there a force causing the trolley to slow down?

   ii. If so, what is this force called?

2. When you have a slope on the runway, the trolley speeds up. Why doesn't it slow down as it does on a level runway?

3. If you put a slope on the runway and the trolley speeds up, which is pulling on the trolley with a larger force, gravity or friction?

4. If, on the other hand, you don't have much of a slope and the trolley gradually slows down, which force is pulling more on the trolley this time?

5. If you adjust the slope on the table so that the trolley moves down the slope with a steady speed, would the forces be the same size or different? What would you expect the dots on a ticker tape to look like?

★ **Information**

To get rid of the effects of friction for our trolley, we can adjust the slope of the runway so that a 'diluted' gravity force just balances the friction force. The trolley will then travel at a steady speed after it has been pushed and the runway will be 'friction compensated'.

■ **What to do**

(a) Collect:
a trolley with a small metal rod
a low voltage supply
a ticker-tape vibrator and tape
about eight small pieces of hardboard
2 elastic cords

(b) Set up the ticker-tape vibrator as in Unit 3.

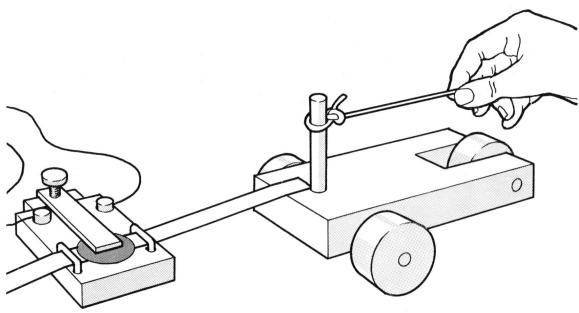

(c) Gradually increase the slope of the runway by sliding bits of hardboard under the end, until the trolley travels at a *steady* speed when it is given a slight push.

(d) Check that the trolley runs down the runway with a steady speed by making a ticker-tape of its motion. (You will have to give it a small push to get it going.)

(e) Fit the metal rod vertically into a hole in the trolley and pass the elastic cord over it.

(f) Repeat the ticker-tape procedure, this time pulling the trolley with a steady force from the elastic cord. Keep it stretched by the same length as the trolley moves.

(g) Make a ticker-tape graph from your tape as in Unit 3.

(h) Now repeat the experiment, this time using two cords instead of one.

▲ What to write

6. When were you using the most force, with two cords or with one?

7. Which graph had the steeper slope?

8. When did the trolley have most acceleration when pulled with one cord or with two?

Extension Work 2
Looking at collisions

★ Information

The particles in a gas move about quickly all over the place. Therefore they very often collide.

We can set up the same sort of collisions on a large scale and study them.

■ What to do

(a) Go to the table with two pucks on it.

(b) Remove one of the pucks and give the other a small push, letting it glide across the table by itself.

▲ What to write

1. Has friction been removed? How do you know?

2. If the table were endlessly long and you were able to remove friction completely, what would happen to the puck after you had given it a push to get it moving?

3. If there are no forces acting on an object, what can you say about its motion?

■ What to do

(c) Place a stationary puck in the centre of the table.

(d) Push another puck straight towards it to make them collide 'head on'.

(Hint: to make this easier to do, line up the two pucks between two half-metre rules.)

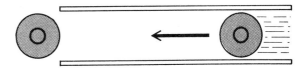

▲ What to write

4. Make drawings to show what happens before and after the collision. Use labels to describe —

  i. how the first puck is moving before the collision

  ii. what happens during the collision

  iii. how the pucks are moving after the collision

Breaking Up is Hard to Do

Unit 1
Heating some substances that are found naturally in our world

★ **Information**

You are going to heat a number of substances. Look carefully at what is left after heating. This is called the *residue*. Notice especially if anything appears to come from the solids as they are heated.

▲ **What to write**

1. What is the purpose of this experiment?
2. What is the name of the solid left after heating?
3. Copy table 1 into your books

■ **What to do**

(a) Heat the materials on a piece of asbestos paper laid on a pipe clay triangle.

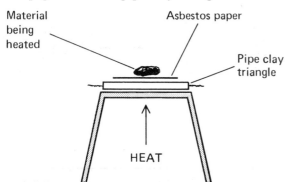

(b) Record your results in the results table.

▲ **What to write**

4. Which materials did not appear to be changed at all by heating?
5. You probably found that many of the substances you heated gave the same coloured residue. What was the colour of the residue?

★ **Information**

All the materials that gave a black residue have at one time been part of a living plant or animal.

The black substance is called carbon. Materials like these, that have a lot of carbon in them, are called *organic substances*.

▲ **What to write**

6. What is the name of the black residue produced in your experiments?
7. What is the name of substances that have a lot of carbon in them?
8. Which of the substances that you heated were organic substances?
9. Which of the following substances contains carbon as one of its constituents?

   Iron   Butter   Petrol   Water

Table 1

| Name of material | Appearance of material | | Any other changes during heating |
	before heating	after heating	

Unit 2
Heating some purified substances in test tubes

★ **Information**

The materials that you will heat in this unit are all obtained from the earth in some way, but they have been purified so that each one only contains one type of substance.

■ **What to do**

(a) Read the Identifying Gases section below. Your teacher will show you how to carry out the tests.

Identifying gases

1. Note the appearance of the gas.
   Its colour and its smell . . .

   (*Note*: to smell a gas, fan the vapour towards your nose with your hand. Be prepared to turn your head away if the gas has an irritating smell.)

2. Often the first gas to come off when heating a substance is *water vapour*.
   If you want to test the vapour for water, hold a piece of *cobalt chloride* paper at the mouth of the tube.

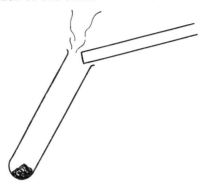

If water is present, cobalt chloride paper goes from blue to pink.

3. Test the gas with indicator paper. Make the indicator paper wet with water and then hold it at the mouth of the test tube.

Note any colour change to the indicator paper.
Red orange or yellow — pH 1 - 6 acid
Blue - green to purple — pH 8 - 14 alkaline

4. Test the gas with a wooden splint.
   i. Test with a <u>glowing</u> splint

Light the splint

Blow out the flame but make sure the splint is still glowing

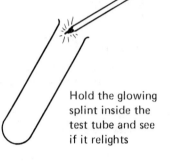

Hold the glowing splint inside the test tube and see if it relights

ii. Test with a <u>burning</u> splint.

Hold the burning splint just inside the mouth of the test tube

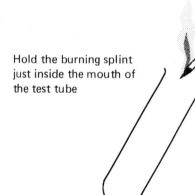

Note to see if the wood splint continues to burn steadily or if it goes out

5. Testing the gas with *lime water* to see if it is carbon dioxide.

Have ready a small ignition tube containing 1 cm depth of clear lime water.

Draw some of the gas you want to test up into a dropping pipette and . . .

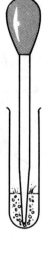

bubble it through the lime water. Repeat it with two more pipettes full of the gas and see if the lime water goes cloudy.

Gas	Form.	Appearance	Smell	Cobalt chloride paper	Indicator paper	Glowing wood splint	Burning wood splint	Lime water
Water	H_2O	"Steam" (colourless)	—	Changes from blue to pink	Neutral	—	Puts the flame out	—
Oxygen	O_2	Colourless	—	—	Neutral	Relights the splint	Continues to burn	—
Hydrogen	H_2	Colourless	—	—	Neutral	—	The gas burns with "pop"	—
Carbon dioxide	CO_2	Colourless	—	—	Slightly acid	—	Puts out the flame	Lime water goes chalky
Chlorine	Cl_2	Yellowish green	Choking smell. Like a swimming pool	—	Acid but bleaches paper white	—	Puts out the flame	—
Nitrogen dioxide	NO_2	Brown	Pungent irritating smell	—	Acid	—	Puts out the flame	—
Sulphur dioxide	SO_2	Colourless— may "flame" in air	Irritating smell leaving "taste in mouth"	—	Acid but may bleach	—	Puts out the flame	—
Ammonia	NH_3	Colourless	Strong pungent smell	—	Alkaline	—	Puts out the flame	—

▲ **What to write**

6. The wood splint is used for testing two different gases. Which two gases?

7. What do we use for testing for water or steam?

8. What colour does cobalt chloride paper go if there is water vapour about?

9. If we dry some pink cobalt chloride paper by holding it over a bunsen flame, what colour will it go?

10. One of the gases you will produce is dangerous if it is breathed in any quantity. The gas is easy to see as it has a brown colour. It usually appears mixed with another gas. When you see this gas come off, test it with a wooden splint, and then don't heat it any more.
What is the name of the brown gas?

■ **What to do**

(b) Collect your gas testing equipment:–
a piece of cobalt chloride paper.
wood splints
a dropping pipette
a small test tube for putting
lime water in.

(c) Also collect a test tube holder, a test tube, a test tube rack.

(d) Put a spatula measure of zinc nitrate in the test tube.

(e) Start to heat the test tube, gently at first.

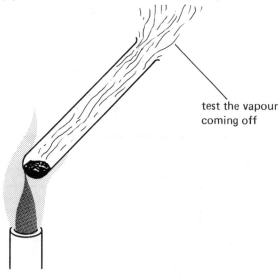

test the vapour coming off

(f) Now heat the substance more strongly until all the water has been driven off. Continue to heat and two more gases will start coming off. Test the gases but do not forget the safety precautions when you see the brown gas.

▲ **What to write**

11. Copy this flow diagram and answer the questions.

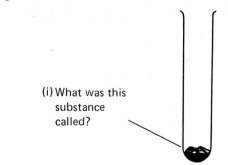

(i) What was this substance called?

(ii) What did it look like?

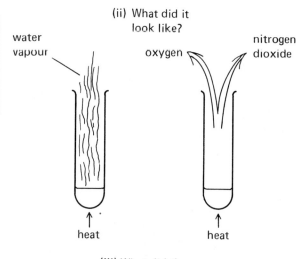

water vapour — oxygen — nitrogen dioxide

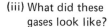

heat — heat

(iii) What did these gases look like?

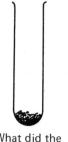

(iv) What did the residue look like?

12. Copy this box chart which shows the chemical changes that have happened.

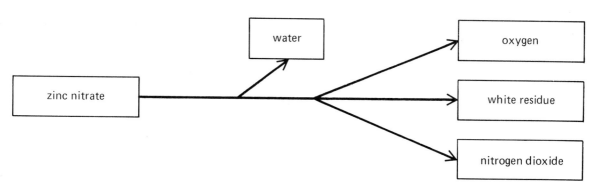

★ **Information**

You should remember from previous work that all substances are made up of tiny particles, too tiny to see even with the most powerful microscope.

Zinc nitrate is a single substance and so it made of one type of particle — zinc nitrate particles.

We could draw a zinc nitrate particle like this:-

In our experiment we changed the zinc nitrate into four different substances.

Each of these four new substances are made up of particles as well.

We could draw:-

a water particle like this:- ⌬

an oxygen particle like this:- ○

a nitrogen dioxide particle like this:-

a particle of white residue like this:- ⌬

▲ **What to write**

13. How many particles has each particle of zinc nitrate been broken up into?

14. Why are the particles of oxygen, nitrogen dioxide etc. drawn much smaller than the particle of zinc nitrate?

Unit 3
Different kinds of reaction

★ **Information**

There are three ways in which particles can behave

   i. Particles can break up into smaller particles.

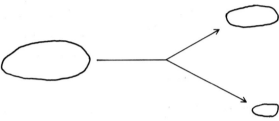

   ii. Particles can join together to make a bigger particle.

   iii. Particles can join together to make a bigger particle which then breaks up into different particles.

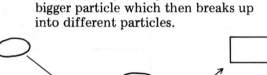

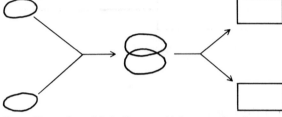

Reactions in which the particles are broken up into smaller particles we call *decomposition reactions*.

▲ **What to write**

1. Decomposition is the name given to a certain type of chemical reaction. Explain, by using a diagram if you like, what the word decomposition means.

2. Copy Table 5 into your book.

Table 5

Name of substance	Colour of substance before heating	Names of any gases produced	Colour of residue	Number of particles
Zinc nitrate	White crystals	Nitrogen dioxide, Water, Oxygen	White	4
Copper carbonate				
Copper oxide				
Copper nitrate				
Zinc carbonate				
Iron carbonate				
Iron nitrate				
Magnesium oxide				

For the column labelled "Number of particles" write down the number of different kinds of particle the substance heated is decomposed into.

■ What to do

(a) Now try heating the substances in the way you heated zinc nitrate.

(b) Fill in your results in the table.

▲ What to write

3. Do all the substances called "_____ carbonate" give off carbon dioxide when heated?

4. Do all the substances called "_____ nitrate" give off nitrogen dioxide when heated?

Unit 4
Breaking up a residue

★ Information

When copper carbonate is heated the following changes take place.

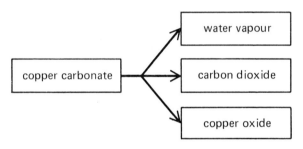

▲ What to write

1. What is the colour of copper carbonate?

2. Look at the results of heating all the compounds of copper. The same residue is left when copper nitrate, copper carbonate, and copper oxide are heated. What is its name?

3. Is the particle of copper carbonate decomposed when heated? If so, how many different kinds of particle is it broken up into?

4. Is the copper oxide decomposed or changed in any way when heated?

★ Information

It is known that a copper oxide particle consists of a copper particle and an oxygen particle joined together.

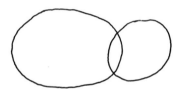

Copper oxide is not decomposed by heating but we are going to try a different way to see if we can decompose the copper oxide particle (by removing the oxygen particle from it).

▲ **What to write**

5. It is known that the copper oxide particle can be split up. What can it be split up into?

6. Can we decompose a copper oxide particle by heating it?

■ **What to do**

(a) Collect:
  a combustion tube (a test tube with a small hole in it)
  a rubber connecting tube
  a bunsen burner
  a clamp and clamp stand

(b) You will need a lab. position where there are 2 gas taps. Put the apparatus together as in the diagram.

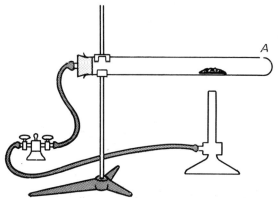

(c) Put 1 spatula measure of copper oxide into the combustion tube.

(d) Turn on the gas flow connected to the combustion tube ¾ of the way, COUNT TO 10, then light the gas coming out of the hole A, and adjust it to give a small steady flame.

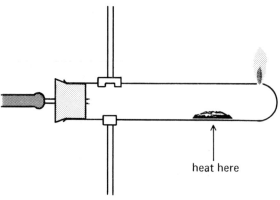

heat here

(e) Now heat the copper oxide for about 10 minutes.

(f) Answer questions 7 and 8 while you are waiting.

▲ **What to write**

7. What is the purpose of this experiment?

8. Draw a diagram of the apparatus and label it.

9. What colour does the copper oxide change to?

10. What do you think the substance left in the tube after the experiment is called?

11. Do you think that the copper oxide particle has been broken up into anything simpler in this experiment?

12. After the experiment there is a different substance in your combustion tube. How can you tell that the copper oxide has been changed into a different substance?

13. If we had weighed the copper oxide before the experiment and then weighed the substance left in the tube after the experiment, which of the following sets of results do you think would be the most likely?

	Set A	Set B	Set C
Mass of black copper oxide	8.0 g	8.0 g	8.0 g
Mass of pink solid left	6.4 g	9.6 g	8.0 g

14. Give reasons for your answer to question 13.

Extension Work 1

■ What to do

(a) Set up the apparatus shown in the diagram.

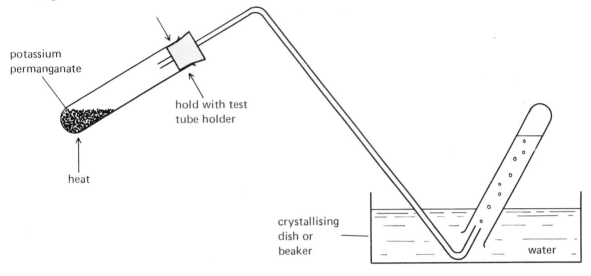

(b) Heat the potassium permanganate (potassium manganate VII) and collect the oxygen gas given off.

(c) Carry out tests on the gas to discover its properties:

Test it with a glowing splint
Is it soluble in water?
What does it look like?
Is it acid or alkaline or neutral?

▲ What to write

1. Draw a diagram of the apparatus you used.

2. Write down a summary of of the gas.

Extension Work 2

■ What to do

(a) Use the same apparatus as in Extension Work 1 but this time put copper carbonate in the tube to be heated.

(b) Heat the copper carbonate and collect the gas that is given off.

(c) Carry out tests on the gas.

What does it look like?
Test it with a glowing or burning splint.
Is it soluble in water?
Is it heavier or lighter than air?
Is it acid, alkaline or neutral?

▲ What to write

1. Describe the tests that you carried out on the carbon dioxide and the results that you obtained.

Atoms

Unit 1
Atoms and elements

★ Information

We have managed to decompose copper carbonate into copper oxide, carbon dioxide and water.

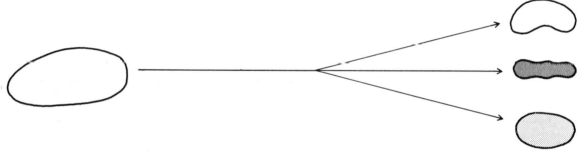

Then, using the combustion tube, the copper oxide particles were broken up when the oxygen was removed from them, and copper particles left.

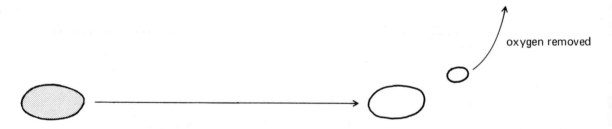

Can we break the copper particles up?

Scientists have many different ways of decomposing or breaking up particles of substances into smaller particles, but whatever is done to copper particles, no method has been found that will break up the copper particle.

70

▲ **What to write**

1. Give the names of three substances you have come across whose particles can be broken up.

2. Which of the following statements is true?

   "Scientists have found only a few ways of breaking up a copper particle."

   "Scientists have only found one way of breaking up a copper particle."

   "Scientists have found no way of breaking up a copper particle."

★ **Information**

In fact, copper is one of a number of substances whose particles cannot be broken up at all. (Their particles seem to be indestructible using ordinary chemical means.)

These indestructible particles, the ones that cannot be broken up, are called *atoms*.

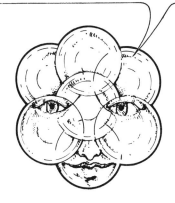

▲ **What to write**

3. What is the name given to a particle that cannot be broken up into a smaller particle?

4. Which particle, that of copper carbonate, or that of copper, is bigger?

5. Which particle, that of copper carbonate, or that of copper, is indestructible?

6. Which particle, that of copper carbonate, or that of copper, is an atom?

★ **Information**

All the atoms of copper are exactly the same as each other.

Other kinds of atom, those of oxygen for instance, are different from those of copper.

▲ What to write

7. What can you say about all the atoms of copper?

8. Are the atoms of oxygen the same as the atoms of copper?

9. What do all atoms have in common?

★ Information

A substance whose particles cannot be broken up, is called an *element*.

So, the particles of an element are all atoms of the same sort.

▲ What to write

10. Give the name of one element.

11. Say if each of the following statements is true or false.

    i. All the particles of one element are the same.

    ii. The particles of an element are all atoms.

    iii. The particles of one element are not always the same as each other.

    iv. The particles of an element are different from the particles of another element.

★ Information

In electricity you used the shorthand symbol for a bulb 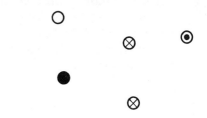 and a battery

In science we often represent an atom by drawing a circle.

The sign ○ means "one atom of an element".

The signs mean "4 atoms of the same element".

We can draw different sorts of atoms by drawing different sorts of circle.

▲ What to write

12. Draw 3 atoms of one element.

13. Draw 3 atoms of one element and 2 atoms of a different element.

14. Draw 5 atoms of one element, mixed with 3 atoms of another element and 2 atoms of a third element.

Unit 2
Elements and compounds

★ **Information**

Atoms of one sort, say oxygen, can become mixed up with atoms of a different sort, say nitrogen.

We could draw that like this.

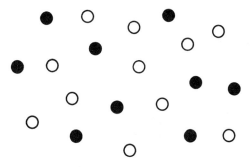

In fact air consists mainly of atoms of nitrogen mixed with atoms of oxygen. (Air can be described as a mixture of elements, or as a mixture of atoms.)

Note: Atoms of gases like oxygen, nitrogen etc. in fact pair up together like this,

but we have drawn them as single atoms for the moment to make it easier.

■ **What to do**

(a) Draw, or collect some plasticine and model 4 atoms of one element.

(b) Now model or draw 2 atoms of one element mixed with 2 atoms of another element.

(c) Now model or draw a mixture of atoms of 3 different elements.

★ **Information**

This is a mixture of hydrogen and oxygen atoms.

But scientists have found that some atoms can become joined together.

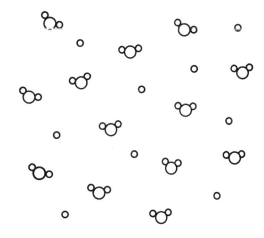

If atoms of different kinds become joined together, the particle formed is called a "particle of a compound".

▲ **What to write**

1. Explain the difference between an atom and a "particle of a compound".

2. Draw what we imagine one particle of the compound water to be like.

★ **Information**

A particle of a compound is made up of at least two different types of atoms joined together.

Such a particle of a compound can be broken up again.

The particle of the compound water

can be broken up into hydrogen atoms

and an oxygen atom.

73

▲ **What to write**

3. Explain the difference between an element and a compound.

4. Make diagrams of the following:—

   i. Some atoms of an element.
   ii. Some atoms of another element.
   iii. Some atoms in a mixture of elements.
   iv. Some particles of a compound.
   v. Some particles of a different compound.
   vi. Some particles of a mixture of an element and a compound.
   vii. Some particles of a mixture of compounds.
   viii. Some particles of a mixture of 2 compounds and 1 element.
   ix. Some particles of a mixture of 2 elements and 1 compound.
   x. Some particles of a mixture of 3 compounds.

5. Atoms of hydrogen and atoms of oxygen can become joined together to form particles of the compound water. Draw a flow diagram to show this happening by:

   i. drawing some hydrogen atoms,
   ii. drawing some oxygen atoms,
   iii. drawing them joined together to form some particles of water.

6. Look at the diagrams below and then answer the questions.

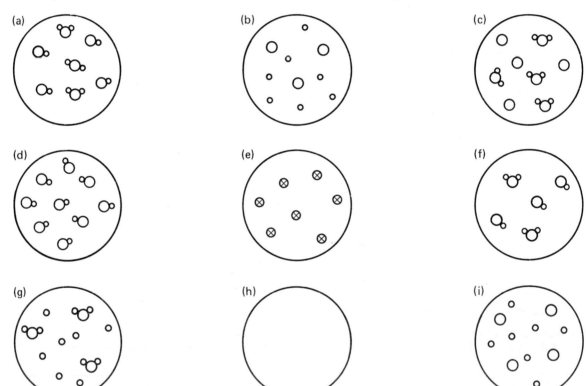

Which diagrams represent

   i. A sample of a pure element?
   ii. Particles of a pure compound?
   iii. A mixture of an element and a compound?
   iv. A mixture of elements?
   v. A mixture of compounds?

7. What does diagram (h) represent?

Elementary My Dear Compound

Unit 1
Making a compound with magnesium and oxygen

■ What to do

(a) Collect:
   a piece of magnesium
   a pair of tongs
   a bunsen burner
   an asbestos mat.

(b) Heat the magnesium in the bunsen flame. When the magnesium starts to burn, *do not look straight at the light*. It will be too bright for your eyes.

(c) Carefully look at the substance you have made. Do not throw it away yet, but keep it for Unit 2.

(d) You will notice that there is some unchanged magnesium together with the compound you have made.

★ Information

When an element burns in air, the atoms of the element join with atoms of oxygen in the air to form a compound.

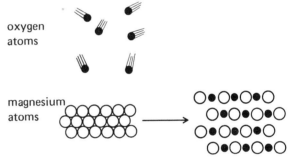

▲ What to write

1. Which element was used to make the compound in this unit?

2. Where did the oxygen come from?

3. Write down and complete this sentence.

   "I have made the compound . . . .".

4. Write down this word equation:

   MAGNESIUM + OXYGEN → MAGNESIUM OXIDE

Unit 2
Comparing properties

▲ **What to write**

1. Copy the magnesium results table shown below.

	Magnesium	Oxygen	Magnesium oxide
Property A. What do the substances look like?			
Property B. Which ones conduct electricity?			
Property C. What happens when they are added to acid?			

■ **What to do**

Property A.

(a) Study the appearance of the elements magnesium and oxygen and the compound magnesium oxide. Fill in property A in your results table.

Property B.

(b) Collect a piece of magnesium.

(c) Test it to see if it will conduct electricity.

(d) Do the same for oxygen, and the compound you have made. Fill in property B in your results table.

Property C.

(e) Collect a bottle of dilute acid
a piece of magnesium
a test tube rack and two test tubes.

(f) Put about 2 cm depth of acid into each test tube.

(g) Into one test tube drop the piece of magnesium. Watch carefully what happens.

(h) Into the second test tube put a small amount of the magnesium oxide you have made. Fill in property C in your results table.

▲ **What to write**

2. How many different sorts of atom are there in magnesium?

3. How many different sorts of atom are there in magnesium oxide?

4. Draw diagrams to show what happens when some magnesium atoms and some oxygen atoms react to make magnesium oxide.

★ **Information**

You have compared the properties of the element magnesium and the properties of the element oxygen with the properties of the compound magnesium oxide.

▲ **What to write**

Which of the following statements is true.

i. The properties of the compound are the same as the properties of the elements that made it.

ii. The properties of the compound are different from the properties of the elements that made it.

iii. The compound has the same properties as one of the elements that made it.

Unit 3
Making a compound with carbon and oxygen

■ What to do

(a) Collect:
 a jar of carbon
 a combustion spoon
 a bunsen burner
 an asbestos mat
 a large test tube of oxygen.

(b) Heat the carbon on the combustion spoon until it glows, and then hold it in the large test tube of oxygen.

(c) When it has finished burning, take out the combustion spoon and replace the bung, so that you can (later) test the gas that is formed.

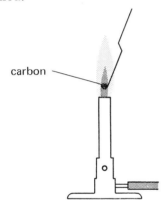

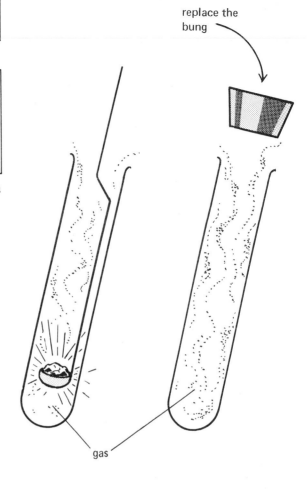

▲ What to write

1. Which elements have been used to make this gas.

2. Write down and complete this sentence:
 "I have made the compound . . . .".

3. How could you test the gas to prove this statement? (See the information section below.)
 Carry out the test, yourself.

4. Write down the word equation for the reaction between carbon and oxygen.

★ Information

Lime water (calcium hydroxide solution) reacts with the gas carbon dioxide and forms the substance calcium carbonate which is insoluble and so makes the liquid go cloudy.

▲ What to write

5. What is the correct name for the gas you have made?

6. How many different sorts of atom are there in carbon?

7. How many different sorts of atom are there in carbon dioxide?

8. You may have called the gas you have made, carbon oxide. This is quite correct, but it is more accurate to call it carbon dioxide. Why do you think this is?
 (Clue: di- before a word, means two.)

9. Draw a diagram showing what happens to the atoms of carbon and oxygen when they make carbon dioxide.

Extension Work 1
Flame tests

★ **Information**

Some metal atoms, when they are heated in a flame, give off light of a particular colour depending upon the metal atoms involved. These colours can be used as a way of identifying those metals.

■ **What to do**

(a) Collect:
a jar of each of the following substances:
sodium chloride
calcium chloride
copper chloride
strontium chloride
barium chloride.

Collect also a bunsen burner
a beaker of distilled water

You will also need a very sharp pencil.

▲ **What to write**
1. Draw a table for your results in this experiment.

Substance	Colour of flame
Sodium chloride	
Calcium chloride	
Copper chloride	
Barium chloride	
Strontium chloride	

■ **What to do**
(b) Dip the tip of the pencil into the water, then into one of the jars of substance so that 1 or 2 crystals are picked up on the tip, and then hold the *tip* of the pencil in the bunsen flame.

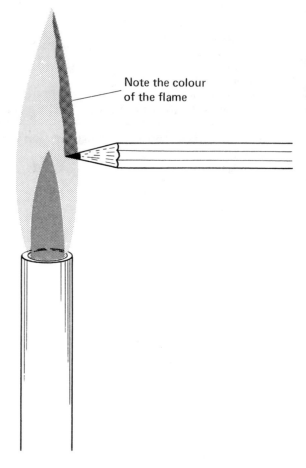

Note the colour of the flame

(c) Note the colour and record your results in your table.

(d) Repeat the experiment with each of the substances but make sure the pencil tip is clean each time.

(e) Collect a jar containing an unknown substance and see if you can use the "flame test" to find out what metal atoms it contains.

Extension Work 2
Another test for metals

★ **Information**

Besides the flame test, some metals can also be recognised by the way their compounds will react with another substance.

In this unit you will try reacting different metal compounds with a solution of sodium hydroxide and then use the results to help to identify an unknown solution.

■ **What to do**

(a) Collect a glass plate and a piece of white paper.

(b) Draw the shape of the glass plate on the paper and then draw a grid of six squares on it.

(c) Collect dropping bottles of
sodium hydroxide
copper sulphate
iron(II) sulphate
iron(III) sulphate
zinc sulphate
magnesium sulphate.

(d) Also collect a bottle containing the solution of a compound of an unknown metal (solution Z).

(e) Place the glass plate on top of the paper.

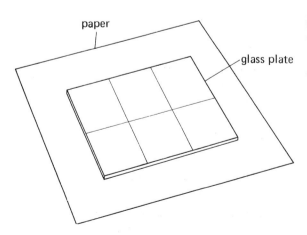

(f) In each square put two drops of sodium hydroxide solution and one drop of one of the metal compound solutions.

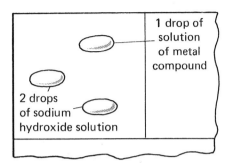

(g) Using a glass rod, mix *one* of the drops of sodium hydroxide solution with one of the drops of metal compound solution.

(h) Note what happens.

(i) Repeat in each square and note what happens in each case.

(j) Now mix the second drop of sodium hydroxide solution in with the mixtures in each square and note if there is any change.

▲ **What to write**

1. Copy out Table 6.

Table 6

Metal compound solution	Sodium hydroxide solution	
	1 drop	2 drops
Copper sulphate		
Iron(II) sulphate		
Iron(III) sulphate		
Zinc sulphate		
Magnesium sulphate		
Solution Z		

2. What metal atom do you think is contained in solution Z?

> **Extension Work 3**
> Examining some oxides of metals and non-metals

★ **Information**

There are many obvious differences between metals (like copper, magnesium, iron) and non-metal elements (like carbon, sulphur and oxygen).

In this unit you are to look at some of the properties of different compounds and then try to see if there is a pattern in the results.

▲ **What to write**

1. Name three differences between metal elements and non-metal elements.
2. Draw Table 7 in your book.

Table 7

Oxide	Appearance of the pure oxide	pH
Copper oxide		
Calcium oxide		
Magnesium oxide		
Iron(III) oxide		
Phosphorus oxide		
Carbon dioxide		
Sulphur dioxide		

■ **What to do**

> (a) Collect:
> a test tube rack
> 6 test tubes
> a spatula
> jars of the first 4 oxides named in your table.

(b) Collect a bottle of diluted universal indicator solution and 1/3 fill each test tube with the solution.

(c) To one test tube, add 1 spatula measure of calcium oxide, shake, and record the pH.

(d) Repeat with the 3 other metal oxides, recording the pH in each case.

(e) Collect a jar of phosphorus oxide and add 1 spatula measure of this compound to a test tube of universal indicator solution. CARE!! DO NOT TOUCH THIS MATERIAL!!!

(f) The sixth test tube we will use to test the oxide carbon dioxide and for this we can use the gas present in our breath. Collect a straw and bubble your breath gently into the universal indicator solution until you notice a change in colour.

(g) To test sulphur dioxide you will need to make some of the gas. To do this collect
a large test tube
a bung
a jar of sulphur
a combustion spoon.

(h) Set alight to a *little* sulphur on the combustion spoon and hold it in the large test tube for about 1 minute. Push the bung into the tube.

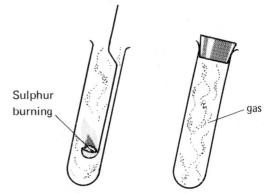

(i) Now add a little universal indicator solution to the tube of sulphur dioxide, replace the bung, shake and note the pH.

▲ **What to write**

3. You should now be able to complete your entries in table 7.
4. What is the one thing all these compounds have in common?
5. What do you notice about the pH of the four metal oxides?
6. What do you notice about the pH of the three non-metal oxides?
7. Assuming that these results are typical for all oxides try to formulate a rule about the pH (or acidity) of metal oxides and non-metal oxides.

Reactivity

**Unit 1
What is reactivity?**

Your teacher will show you how some different metals react with water. The diagram below shows what happens when sodium is put into water:

▲ What to write
1. Did all the metals react with water?
2. Try to put these metals in an order of reactivity, most reactive first.

   sodium, magnesium, copper, potassium, iron, lithium, zinc.

★ Information
Let us now look more closely at the elements which seem to have similar reactivity:

   copper, iron, magnesium, zinc.

Scientists represent different atoms by using different signs.

Magnesium's sign is	Mg
Copper's sign is	Cu
Iron's sign is	Fe
Zinc's sign is	Zn

▲ What to write
3. A magnesium atom could be drawn like this:

Now draw symbols for atoms of iron, copper and zinc.

81

> Unit 2
> Comparing the reactivity of metals with acid

■ What to do

(a) Collect:
4 test tubes
a wooden splint
containers of copper
                    iron
                    zinc
                    magnesium
a bottle of hydrogen chloride solution.

★ Information

Hydrogen chloride solution is an acid. It is usually called hydrochloric acid.

■ What to do

(b) Half fill each test tube with the acid.

(c) Add a spatula measure of zinc to the acid. If there is a gas given off, test it like this.

Put your thumb over the mouth of the test tube

After 30 secs, or when you feel the pressure of gas on your thumb . . .

. . . hold a burning splint to the gas

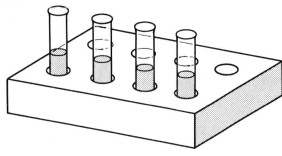

(d) Throw away the contents of the test tube into the "waste chemical container".

▲ What to write
1. Copy Table 8 into your books.

Table 8

Metal	Speed of reaction	Name of gas given off
Zinc		
Copper		
Iron		
Magnesium		

Write the answers to questions 2 and 3 in the table.

2. For zinc, say if the reaction is "very fast", "fast", "slow" or "no reaction".
3. What is the name of the gas given off?

■ What to do

(e) Now try reacting the other metals with the acid.

▲ What to write
4. Complete the table.
5. Write down the metals in order of their reactivity, starting with the most reactive.

★ Information

Hydrogen chloride is a compound of two atoms.

   Hydrogen and chlorine joined together

▲ What to write
6. There are two types of atom in hydrogen chloride. What are their names?

★ Information

In this experiment the metals react with the acid.

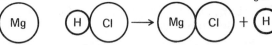

▲ What to write
7. Draw diagrams like the one above to show what happened when each of the following metals was put into hydrochloric acid.

zinc, iron, copper.

You will learn later that the reaction is a bit more complicated than this — for the moment, don't worry!

Unit 3
Where have the metals gone?

■ **What to do**

(a) Collect an evaporating basin.

(b) Pour one test tube full of hydrochloric acid into the basin.

(c) Add a piece of magnesium ribbon, roughly 5 cm long, to the acid.

(d) When the metal has finished reacting, evaporate the liquid.

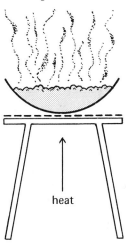

heat

▲ **What to write**

1. Describe what the solid left in the evaporating basin looks like.

2. Suggest a name for the solid left in the basin.

3. What gas is given off as the magnesium reacts with the acid?

4. What substance went into the air when you evaporated the liquid?

5. Is the solid left in the basin, magnesium?

6. Are the magnesium atoms still in the evaporating basin?

7. How many different kinds of atom are there in the solid you have made?

8. What are the names of the different kinds of atom?

9. What would the solid look like if the chlorine atoms were removed in some way?

Unit 4
Making copper join with chlorine

★ **Information**

You have found that copper is not reactive enough to push the hydrogen atoms out of the hydrogen chloride.

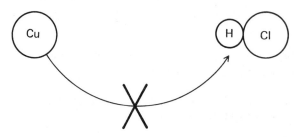

But copper can bring along reinforcements to help remove the hydrogen atoms from the hydrogen chloride.

Oxygen is good at dragging away hydrogen atoms.

Let's try to see if copper, with the help of oxygen, can remove the hydrogen from the hydrogen chloride.

▲ **What to write**

1. We will need a compound containing copper and oxygen atoms. What is the name of the compound formed when copper and oxygen atoms join together?

2. Explain in your own words what you are going to try to do in this experiment.

■ **What to do**

(a) Collect
   a test tube
   a container of copper oxide
   a bottle of hydrochloric acid.

(b) ½ fill the test tube with hydrochloric acid.

(c) Put ½ a spatula of copper oxide into the acid.

(d) Shake the mixture and hold it up to the light for about 2 minutes.

▲ **What to write**

3. Describe what you saw happening in the test tube.

4. Copper chloride solution is blue-green in colour. What evidence is there to suggest that you have made copper chloride solution?

★ **Information**

You will have realised that you have made a solution of copper chloride.

The copper atoms could not push the hydrogen from the hydrochloric acid, but the oxygen part of the copper oxide has dragged away the hydrogen atoms:

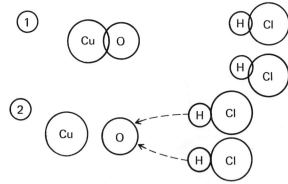

Oxygen drags hydrogen off
. . . leaving the copper and chlorine atoms free to make copper chloride.

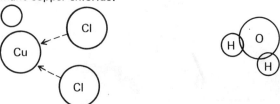

▲ **What to write**

5. What happens to the hydrogen atoms in the hydrochloric acid when the acid reacts with copper oxide?

6. What happens to the oxygen atoms of the copper oxide when it reacts with the acid?

7. Draw a particle that is formed when hydrogen and oxygen atoms join together.

8. What is the name of this compound?

9. Scientists often use equations to show what is happening in a reaction. Try to complete this word equation.

copper oxide + hydrogen chloride

= .... + ...

Unit 5
Can we get the metals back from the solutions?

■ **What to do**

(a) Collect:
a petri dish
a piece of magnesium
a bottle of copper chloride solution.

(b) Pour a little of the copper chloride solution into the petri dish.

(c) Put a piece of magnesium into the solution and leave it.

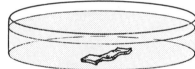

(d) Now read the information section and answer question 1.

★ **Information**

We know that there are copper atoms in copper chloride solution. We are now trying to remove the copper atoms from the chlorine atoms.

What is happening in this reaction is rather like boys playing football. One of the metals, copper, is starting off with the ball,

and is challenged for it by magnesium. The more reactive metal will win.

▲ **What to write**

1. Which metal do you think will finish up with the chlorine?

■ **What to do**

(e) Now look at the petri dish and wait to see if your prediction is correct.

▲ **What to write**

2. What happens to the magnesium metal in the copper chloride solution? Has any copper metal been produced?

3. Is the copper losing possession of the chlorine atoms?

4. What is happening to the magnesium atoms?

5. Draw particle diagrams to show what you think is happening in the reaction.

6. Copy Table 9 into your books.

Table 9

	Copper chloride	Iron chloride	Magnesium chloride	Zinc chloride
Copper				
Iron				
Magnesium				
Zinc				

■ What to do

(f) Carry out experiments to get evidence to help you fill in Table 9. You may like to combine with other groups in order to get a complete set of results quickly.

Here are the four experiments that must be set up.

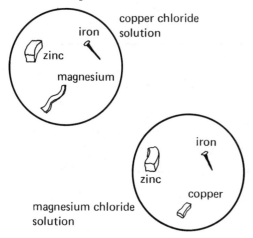

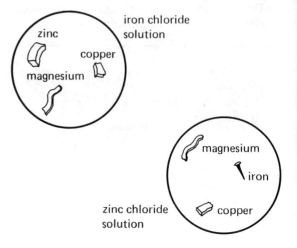

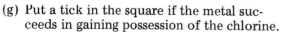

(g) Put a tick in the square if the metal succeeds in gaining possession of the chlorine.

▲ What to write

7. Which metal gains possession of the chlorine in the most number of cases?

8. Which metal loses possession of the chlorine the most frequently?

9. On the basis of the results, try to put the metals in an order of reactivity.

10. Is the order of reactivity the same as the one you worked out for the reactivity of the metals with acid?

Extension Work 1
Using a blow pipe

★ **Information**

Most metals in the earth's crust are present in compounds, like iron oxide or copper carbonate. In this experiment you are going to try to obtain lead from lead oxide.

■ **What to do**

(a) Collect:
   a blow pipe and carbon block
   a bottle of lead oxide
   a bunsen burner.

(b) With a knife, scrape a small hole in the surface of the carbon block.

(c) Pack into the hole a little lead oxide and add to it a drop (or two) of water so that it forms a paste.

(d) Now use the blow pipe and bunsen burner to heat the lead oxide as hard as you can. Look carefully at the diagrams below to see how this is done.

▲ **What to write**

1. What did the lead oxide look like?

2. What did the lead metal look like?

3. If lead oxide is heated strongly without the carbon block it does not change into lead. Why do you think it is important for the lead to be heated on the carbon block?

4. Write an equation to explain your answer to question 3.

Extension Work 2
The blast furnace — making iron

■ **What to do**

Look at the charts and posters that give information about how iron ore is changed into iron, and then answer the following questions.

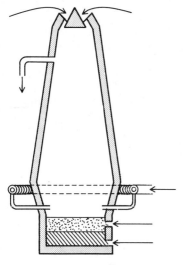

▲ **What to write**

1. Using a full page of your book, copy this outline diagram of the blast furnace.
2. Now carefully label on your diagram:
    i. the 3 raw materials that are put into the top of the furnace;
    ii. what is blown into the bottom of the furnace,
    iii. the approximate temperature of the furnace,
    iv. what the two layers of liquid at the bottom of the furnace are;
    v. the names of any waste gases that escape from the top of the furnace.
3. Explain why you think the iron oxide (iron ore) is changed into iron by this process.
4. What is the purpose of the limestone in the furnace?

Extension Work 3
Methods of extracting metals

★ **Information**

If we make a list of some metals, and place them in order of their reactivity, we can also add carbon to the list.

    sodium
    calcium
    magnesium
    aluminium
    CARBON
    zinc
    iron
    tin
    lead
    copper
    mercury
    silver
    gold

▲ **What to write**

1. The blast furnace method of heating a metal ore with coke in a blast of hot air is used for other metals besides iron. With which metal ores do you think the blast furnace method will be successful?
2. Name some metal ores that the blast furnace could not be used to extract.
3. As the earth cooled, all the metal atoms began to compete with each other to join with the available non-metal atoms like oxygen, sulphur, carbon etc. Some metals however failed to join with other atoms and remained in the earth's crust unreacted. Which metals in the list above do you think may have failed to form compounds?
4. Find out about the method used to obtain those metals that cannot be extracted using the blast furnace method.

12

Rates of Reaction

Unit 1
The effect of heat on a reaction

★ **Information**

You probably know that when health salts are put in cold water a reaction occurs and a gas is given off.

In this experiment you will find out if the temperature of the water changes the speed at which the health salts react.

▲ **What to write**

1. Explain what you are trying to find out in this experiment.

■ **What to do**

(a) Collect a 100 cm³ beaker and half fill it with water.

(b) Collect another 100 cm³ beaker and half fill it with hot water.

(c) Collect a container of health salts.

(d) Now add two spatula measures of health salts to each beaker.

▲ **What to write**

2. What happens when you put the health salts in water?

3. In which beaker did the health salts react the fastest?

4. Health salt particles react with water particles when they bump into each other. Particles go faster when they are hotter. Now try to explain why the reaction was quicker in warm water.

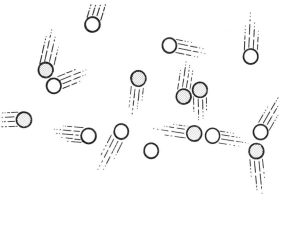

**Unit 2
Reacting zinc with acid at different temperatures**

★ **Information**

You may remember that when zinc is put into dilute hydrochloric acid, the zinc reacts and bubbles of hydrogen gas are formed.

▲ **What to write**

1. Work out a way to find out if the reaction between zinc and dilute hydrochloric acid gets faster at higher temperatures. The following materials and apparatus are available in the laboratory:

    cold dilute hydrochloric acid
    warm dilute hydrochloric acid
    zinc pieces
    test tubes.

    Draw diagrams or write a plan to explain how you are going to do this.

■ **What to do**

(a) Having had your ideas checked by your teacher, carry out the experiment.

▲ **What to write**

2. How can you tell how fast the zinc is reacting?
3. Does the reaction go faster at high or low temperatures?
4. From the evidence so far, which of the following statements seems to be correct?

    i. Temperature has no effect on the speed of a reaction.
    ii. Temperature increases the speed of all reactions.
    iii. All the reactions we have tested have been speeded up by higher temperatures.

Unit 3
Catalysts: another way to speed up a reaction

★ Information

So far you have discovered that substances react faster at higher temperatures. In this experiment you will be finding out about another way of speeding up a reaction.

■ What to do

(a) Collect a test tube rack and two test tubes.

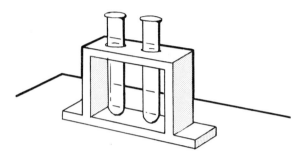

(b) Pour about 4 cm depth of dilute hydrochloric acid into each test tube.

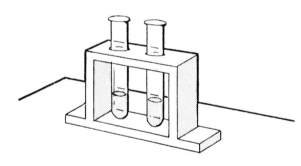

(c) Add 3 pieces of copper to *one* of the test tubes.

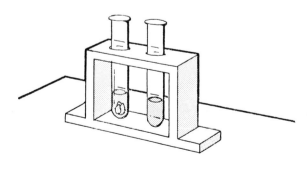

(d) See if there is any reaction between the acid and the copper.

(e) Now add one spatula measure of zinc shavings to each of the test tubes.

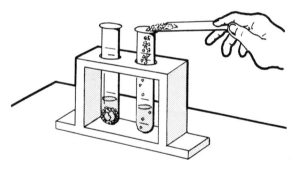

(f) Compare the speed of the reaction in the two test tubes.

▲ What to write

1. Does copper react with the dilute hydrochloric acid?

2. Does zinc react with the dilute hydrochloric acid?

3. Does zinc react faster when copper is present in the test tube?

4. You have found another way of speeding up the reaction of zinc with dilute hydrochloric acid. Which of the following statements best explains how the reaction was speeded up?

   i. Copper reacts with acid to produce a gas and so speeds up the reaction.

   ii. Zinc reacts with acid to produce a gas. This gas then makes the rest of the zinc and acid react faster.

   iii. Although copper does not react with hydrochloric acid, it somehow seems able to speed up the reaction between zinc and acid.

★ Information

Substances that can speed up chemical reactions without themselves being used up are called *catalysts*.

▲ What to write

5. What is the name of a substance that can speed up a chemical reaction without itself being used up?

6. In the Unit 3 reaction, which substance was the catalyst?

7. Catalysts are used a lot in industry. Give two reasons why they are worth the money, even though they are sometimes very costly to buy.

Extension Work 1
How does concentration affect the speed of reaction?

★ Information

So far in this work you have seen how the speed of reaction is affected by catalysts and temperature. In these extension units you will be trying to find out if any other factors affect the speed of reaction.

■ What to do

(a) Collect:
two 100 cm³ beakers
a test tube
a bottle of dilute acid
a bottle of calcium carbonate

(b) Into one beaker pour a test tube full of water.

(c) Now to each beaker add 1 test tube full of the dilute acid.

(d) Get two spatula measures of calcium carbonate ready to pour into the two beakers.

(e) At the same moment, add the calcium carbonate to each beaker.

(f) Observe the reaction.

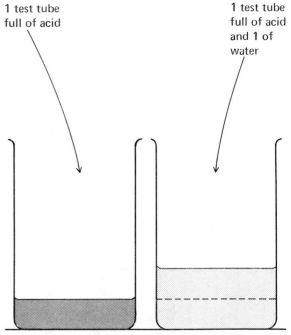

1 test tube full of acid

1 test tube full of acid and 1 of water

▲ What to write

1. What is the name of the gas being evolved in this reaction?

2. How would you test the gas to prove your answer to question 1?

3. Was there any difference in the speed of the reaction in each beaker? If so, which was the fastest?

4. In your own words, say how you think the concentration of the reactants affects the speed of the reaction.

5. By using diagrams, try to explain why the concentration affects the speed of the reaction. Try to make your explanation as full as you can.

   Here is a diagram to give you a start.

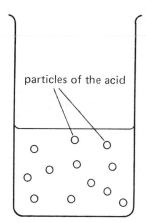

particles of the acid

Extension Work 2
Reacting acid with marble chips of different sizes

■ What to do

(a) Collect:
two 100 cm³ conical flasks
a test tube rack
two test tubes
a bottle of dilute acid.

(b) You will find that there are two sizes of marble chips available. Weigh out 10 g of the small chips and 10 g of the large chips.

★ Information

Marble, limestone, and chalk are all different forms of the same material, calcium carbonate.

(c) Put the two portions of marble chips into the two conical flasks.

(d) Fill the two test tubes with dilute acid.

(e) At the same moment, add the acid to the marble chips.

(f) Observe the speed of reaction.

▲ What to write

1. What is the name of the gas that is evolved in the reaction?

2. Was there any difference in the speed of the reaction in each flask? If so which was the fastest?

3. By using diagrams, try to explain why one reaction went at a different speed from the other. (Remember the same quantities were used in each case.)

   Here is a diagram to help you start.

10 g of large marble chips
could be represented by . . .

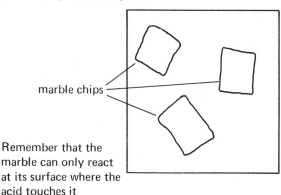

marble chips

Remember that the marble can only react at its surface where the acid touches it

13

Get a Move On

Unit 1
Evaporation: A reminder

★ **Information**

If a drop of water (or any liquid) is left open in a room, it eventually disappears. It is said to have *evaporated*.

In previous work, we have tried to explain this by thinking about the movement of the particles of the liquid.

▲ **What to write**

1. Look at the diagram showing particles moving in a liquid. Are all the particles moving at the same speed?

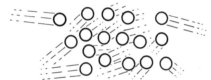

★ **Information**

A drop of water on a table top, or a rain drop hitting a window pane tends to stay together as a drop. Scientists explain this by saying that there must be forces holding the particles together in the drop.

▲ **What to write**

2. Draw a number of particles of liquid, like those on the right, and then draw lines between the particles to show the forces that act between them. We call these forces "holding forces".

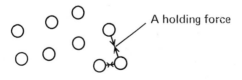

3. Look at the lines of force that you have drawn and the directions of these.

   i. Do any of them act to push the particles apart?

   ii. Do all of them act to pull the particles closer together?

4. When a liquid evaporates, some of the particles are escaping from the liquid. Only certain particles are able to escape. Which ones? Can you explain why?

5. A hot liquid evaporates faster than a cold liquid. What is different about the particles of the hot liquid that causes this to be true?

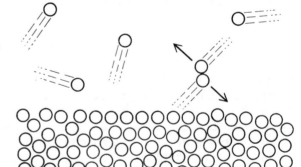

▲ **What to write**

6. After looking at the picture, write down as many differences as you can see between the particles in the liquid and those in the

gas. Think about: how fast they move, how much they are spread out, how often they collide.

7. If it were possible to look closely at a particle from the gas and a particle from the liquid, would there be any difference between them?

8. Why do you think a liquid will not evaporate completely from a stoppered bottle?

9. If a fast-moving liquid particle escapes from the surface of a liquid, what could happen to that particle that would cause it to be knocked back into the liquid?

The next section is about what happens when liquids boil.

★ Information

Imagine a section of the surface between a liquid and the air.

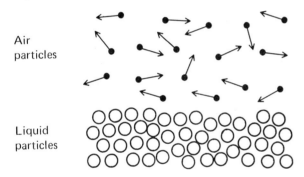

▲ What to write

10. What is the name of the pressure that acts downwards on the surface between the liquid and the air?

11. What is the name of the pressure that acts upwards on the surface?

12. Explain what causes air pressure.

★ Information

There are two things that tend to stop a liquid particle from escaping from a liquid. One is the holding force between the particles of the liquid, and the other is that a particle has got to escape against the *air pressure* acting on the surface of the liquid. (In other words, to escape, a liquid particle must not be hit by a fast moving air particle as it leaves the liquid.)

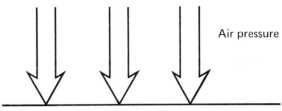

▲ What to write

13. There are two things that discourage a liquid particle from evaporating. What are they?

14. What do you think will happen to the strength of the vapour pressure as the liquid is heated?

15. Explain why you think the air pressure will or will not change when a liquid is heated.

★ Information

As a liquid is heated, the holding forces between the particles do not alter. The particles just move faster because of the extra energy they have received.

As a liquid is heated the vapour pressure increases.
As a liquid is heated the air pressure does not change.

▲ What to write

16. At its boiling point, a liquid begins to evaporate rapidly. Try to explain why this is.

17. At its boiling point, bubbles of vapour are formed inside a liquid. Try to explain why this is.

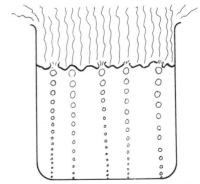

18. On top of Mount Everest water boils at a temperature well below 100°C. Explain why it should boil at such a low temperature.

Unit 2
Freezing

▲ **What to write**

1. Collect a piece of graph paper and mark on it the following axes and scales.

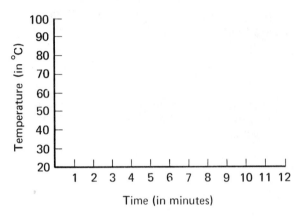

■ **What to do**

> (a) Collect:
> a tripod, gauze, and asbestos mat
> a clamp and clamp stand
> a Bunsen burner
> a tube containing a substance called naphthalene
> a 250 cm³ beaker
> a thermometer

(b) Set up the apparatus as shown in the diagram.

(c) Heat the water until it nearly boils and all the naphthalene has melted.

(d) Take the test tube from the water and fix it in the clamp.

(e) Note the temperature of the naphthalene. Record the temperature on your graph at time = 0.

(f) Now, occasionally stirring, note the temperature of the naphthalene every minute for 12 minutes. Record the temperatures on your graph.

▲ **What to write**

2. What is the freezing temperature of naphthalene?

3. When all the naphthalene cools to its freezing temperature, does all the naphthalene freeze suddenly or over a period of time?

4. What is happening to the particles of the naphthalene as they change from being in a liquid state to a solid state?

5. What is happening to the speed of movement of the particles of naphthalene as the naphthalene cools.

6. What is the name for the forces that cause the naphthalene particles to cling together and form the solid?

7. Why do the naphthalene particles cling together at the freezing point, but do not do so at a temperature just above the freezing point?

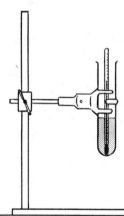

Unit 3
How particles move in a solid

★ **Information**

In a solid, the particles are moving but are held in place by the holding forces between the particles. The result is that they are constantly vibrating.

■ **What to do**

We can get an idea of the way particles move in a solid if we make a "flick book"

(a) Draw 4 circles about this size in the right hand corner of the right hand page of your exercise book.

(b) Repeat this drawing of the 4 circles on each page of your exercise book. Do it quickly and do not worry if they are not exactly the same size or in the same place each time.

(c) Now hold the pages so that you can flick the pages rapidly and look at the 4 circles all the time.

▲ **What to write**

1. Use your own words to describe the way the circles have appeared to move.

2. The way the circles moved is very like the way the particles are moving in a solid. From the following list of words and phrases pick out the one that describes this movement best:

   sliding,
   vibrating,
   tumbling over each other,
   flying about,
   bumping into each other and flying off in different directions, like "dodgems" in a fair, like people in a football crowd just before the final whistle.

3. From the same list, pick out the word or words that best describe the way you imagine particles in a liquid are moving.

4. Pick out the word or words that describe the way you think particles in a gas move.

Unit 4
Melting points

★ **Information**

Why do different solids melt at different temperatures? This is the question that this unit tries to answer.

▲ **What to write**

The chart on the right shows the temperatures at which different substances melt/freeze. Use it to answer the following questions.

1. What substance freezes at 0°C?

2. Which substances freeze at a temperature below 0°C?

3. The temperature of a hot bunsen flame is about 600°C. Which substance should it just be possible to melt?

4. Could common salt (sodium chloride) be melted using a bunsen burner?

5. Arrange the following six substances into two groups.

   Group A — Substances that need a lot of heat to melt them.

   Group B — Substances that need a little heat to melt them.

   The six substances to group are:

   magnesium oxide
   naphthalene
   candle wax
   iron
   ice
   sodium chloride.

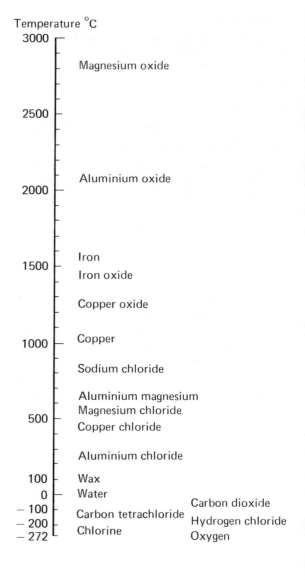

Unit 5
How atoms are joined together

★ **Information**

We believe that compounds are made up of definite particles.

e.g.

Mg——O   a magnesium oxide particle — one atom of magnesium joined to one atom of oxygen

   a water particle — two atoms of hydrogen joined to one atom of oxygen

▲ **What to write**

1. In solids, do you think there must be a force holding these particles together?

   If your answer is "Yes" go to question 3.

   If your answer is "No" go to question 2.

2. If there is no holding force between the particles, why do you think a solid can keep its shape and not fall apart?

   Write down your answer and show it to your teacher.

3. We call this force the "holding force". In which group of substances would you expect this holding force to be very strong, Group A or Group B?

4. By a process called "X-ray crystallography", scientists have found that magnesium oxide crystals are made up of atoms arranged like this:

   Mg   O   Mg

   O   Mg   O

   Mg   O   Mg

Which atom of oxygen do you think the magnesium atom in the middle is joined to?

In fact the correct answer is that the magnesium atom is joined to all four oxygen atoms:

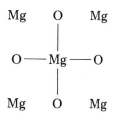

5. Now copy out and draw in all the holding forces in this structure:

   Mg   O   Mg   O   Mg   O

   O   Mg   O   Mg   O   Mg

   Mh   O   Mg   O   Mg   O

6. Explain why you think magnesium oxide is so difficult to melt.

7. The atoms in magnesium oxide are said to be arranged in a *giant lattice structure*.

8. In which group of substances, Group A or Group B, would you expect the holding forces to be fairly weak?

99

9. Scientists have found that the atoms in ice are arranged like this.

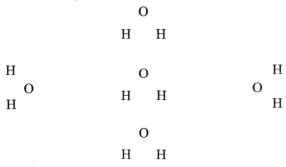

Which atoms of hydrogen do you think the oxygen atom in the middle is joined to?

10. The particles of water $\overset{O}{\underset{H\ H}{\wedge}}$ are called water molecules. How many molecules of water are there in the diagram?

A molecule is the name given to a particle made up of a small number of atoms joined tightly together. The holding forces between the atoms in a molecule are very strong but the holding forces between different molecules are fairly weak. Copy the diagram showing the atoms in ice and mark in, in different colours:

i. the strong holding forces;

ii. the fairly weak holding forces.

12. Which of the following statements (a) – (g) are true about giant lattice structures?

(a) The holding forces between the atoms are all equally strong.
(b) Some of the holding forces are very strong and some are weak.
(c) The melting points are high.
(d) The melting points are low.
(e) The substance is made up of a network of atoms in which each atom is joined equally to all those surrounding it.
(f) When it melts each atom becomes free to move.

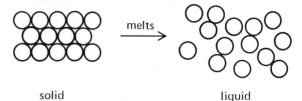

solid   liquid

(g) When it melts, small groups of atoms become free to move but the atoms within the groups are still held together.

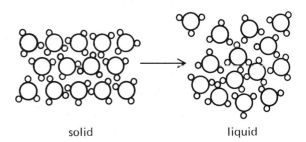

solid   liquid

13. Which of the statements (a)–(g) are true about substances made up of molecules (molecular structures)?

14. The particles that a molecular structure breaks up into when it melts are different from the particles a giant structure breaks up into when it melts. What is the difference?

Unit 6
Testing solids

■ What to do

(a) There are a number of substances available for this unit. Carry out tests on them to find out if they have a molecular structure or a giant structure.

▲ What to write

1. Make a report of the results of your tests.

Extension Work 1
Electrolysis

■ What to do

(a) Collect the apparatus necessary to set up the sort of apparatus shown in the diagram.

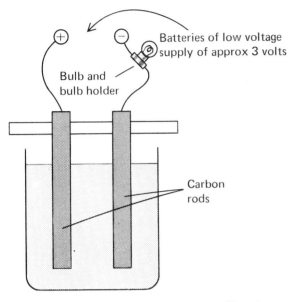

(b) Carry out an experiment to see if water conducts electricity.

(c) Now dissolve two spatula measures of copper chloride in the water.

(d) Carry out an experiment to see if copper chloride solution conducts electricity.

(e) Pass the electric current through the copper chloride solutions and try to find out what is happening at the + electrode and what is happening at the − electrode.

▲ What to write

1. Does water conduct electricity in your experiment?

2. Does copper chloride solution conduct electricity?

3. What is formed at the + electrode?

4. What is formed at the − electrode?

5. What atoms are in copper chloride?

6. What is happening to the atoms in copper chloride in this experiment?

★ Information

There are two types of electric charge, plus (+) and minus (−).
+ electric charges attract − electric charges.
+ electric charges repel other + charges.
− electric charges repel other − charges.

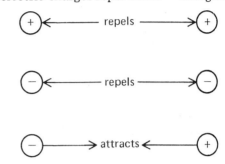

▲ What to write

7. Scientists believe that atoms in compounds like copper chloride have charges on them. What sort of charge (+ or −) has the copper atoms in copper chloride?
What sort of charge have the chlorine atoms in copper chloride?

8. The copper and chlorine atoms in (solid) copper chloride can be drawn like this. Copy them and mark in on the diagram the electric charges on the atoms.

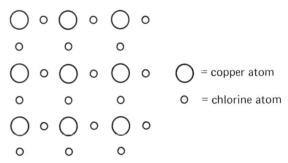

9. Try to explain what you think might be holding the atoms together in the solid.

■ What to do

(f) Use your apparatus to find out if other substances can be split up by electricity. Your teacher will tell you which ones to try.

**Extension Work 2
Metals**

■ What to do

(a) Look at the display of metals.

▲ What to write

1. From what you have seen, would you say that metals have a crystalline structure?

2. What does this tell you about the way atoms are arranged in metals?

**Extension Work 3
Growing crystals**

■ What to do

Try growing your own crystals.

Ask your teacher to tell you how to do this.

14

Merging Molecules

Unit 1
Predicting what will happen when sulphur is heated

★ **Information**

"Predicting" is trying to say what the results of an experiment will be before you actually do it.

■ **What to do**

(a) Look at the samples of sulphur.

▲ **What to write**

1. The sulphur is a solid. Does this mean that the atoms in sulphur are close together or far apart?

2. As a sulphur crystal has a regular shape, which of the following do you think is true?

    i. The atoms in sulphur are probably arranged in a regular pattern.

    ii. The atoms in sulphur are probably jumbled together any-old-how.

★ **Information**

In sulphur, the atoms are linked together in groups of 8 atoms in the form of a "ring".

Look at the diagram showing how, in the crystal of sulphur, the bulges in one ring of sulphur atoms fit into the hollows of its neighbours.

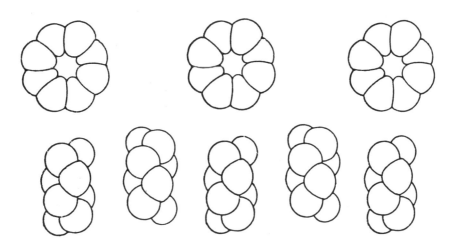

103

▲ **What to write**

3. From the diagram, do you think that sulphur has a molecular or a giant structure?

4. What sort of melting point do you think sulphur will have — high or low?

★ **Information**

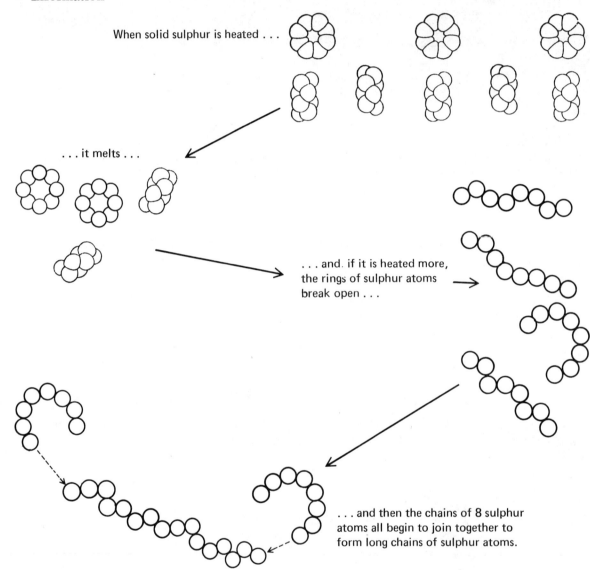

▲ **What to write**

5. In the solid state, are the atoms of sulphur arranged in rings of 8 sulphur atoms or in chains of 8 atoms?

6. When the sulphur is melting, are the atoms of sulphur arranged in rings of 8 sulphur atoms or in chains of 8 sulphur atoms?

7. When the sulphur is heated more, what happens to the rings of sulphur atoms?

8. When sulphur first melts, it is a runny liquid. What do you think will happen to this runnyness when the sulphur atoms begin to join together?

★ Information

When the sulphur cools, the long chains of sulphur atoms break up into chains of 8 atoms again; these then curl up to reform the rings, making the hard, yellow sulphur we started with.

But, if we cool the sulphur *very* quickly, the atoms do not have time to do this, so that they are still joined together in long chains. This is called "plastic sulphur".

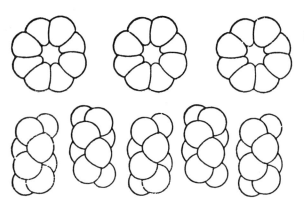

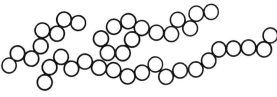

▲ What to write

9. Explain the difference in structure between plastic sulphur and yellow sulphur.

10. Is it true to say that the only difference between the plastic sulphur and the yellow sulphur is the arrangement of the atoms?

Unit 2
Heating sulphur

★ **Information**

This is an experiment that needs to be done carefully if you are to see all that is happening.

■ **What to do**

(a) Collect:
a bunsen burner
an asbestos mat
a 250 cm³ beaker.

(b) About ¾ fill the beaker with water.

(c) Now collect a test tube with some sulphur in it.

(d) CARE.
Heat the sulphur gently and keep shaking it until the sulphur has *partly* melted.

(e) When you have noticed how runny the liquid is, continue to heat it and watch carefully all the changes that take place.

(f) When the sulphur starts to boil, pour it quickly into the beaker of water.
CAREFUL! IT MAY CATCH FIRE.
IF IT DOES, PUT A DAMP CLOTH RIGHT OVER THE TOP OF THE FLAME.

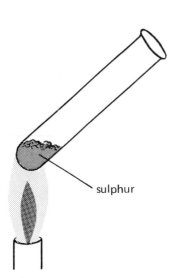

(g) As soon as the sulphur is cool, take it out of the water and examine it carefully and then keep it until the next lesson.

▲ **What to write**

1. Complete these statements.

   i. The melting point of sulphur is .... *low/high*

   ii. When sulphur first melts it is .... *runny/thick*.

2. What happens to the runnyness of the melted sulphur as it is heated more?

3. How could you tell the point in this experiment when the chains of sulphur atoms are joining together?

4. What did the sulphur feel like after the boiling sulphur has suddenly been cooled in cold water?

5. Plastic sulphur looks and feels different from the sulphur you started with.

   i. Are both these forms of sulphur, elements?

   ii. Are they the same element?

   iii. Do both forms of sulphur have the same kinds of atoms?

   iv. Do both forms of sulphur have their atoms arranged in the same way?

6. What is it that makes plastic sulphur (the sulphur you made by heating ordinary sulphur and then cooling it suddenly) have different properties from ordinary sulphur? (Look at your answers to question 5 to help you.)

7. After about a week, plastic sulphur becomes hard and yellow once more. What do you think must have happened to the arrangement of the atoms in the sulphur?

Unit 3
Making Nylon: A chemical rope trick

★ **Information**

Nylon is made from two different substances: *adipyl chloride* and *diamino hexane*.

As these are both difficult names to remember we will refer to them by the abbreviations Ad.C and Di.H.

They are both long thin molecules ...

 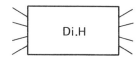

and each molecule has a special reactive group of atoms at each end.

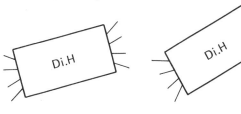

The special thing about these groups of atoms is that they cause the molecules of one substance to join with the molecules of the other substance.

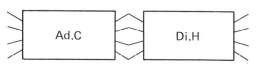

But the molecules of the same substance will not join together.

■ **What to do**

(a) Collect a tray labelled "Nylon Model".

(b) Each block in the tray represents a molecule of either Ad.C or Di.H. Make sure that they are all separate, spread out, and have their painted faces upwards.

(c) Now shake the tray gently from side to side so that the "molecules" start to slide about.
Watch what happens!

▲ **What to write**

1. Draw what your tray of models looked like *before* you started to shake them.

2. Now draw what the models look like after you have shaken the tray from side to side.

3. Explain what happens to the molecules of Ad.C. and Di.H. when they collide together.

4. Nylon is formed when thousands of molecules of Di.H. and Ad.C join together. Would you expect the giant molecules of nylon to be like:-
   long chains
   big balls
   large rings
   some other shape?

★ **Information**

When we make nylon in the laboratory we use two solutions.

   Di.H dissolved in water.

   Ad.C dissolved in tetrachloromethane.

These two solutions do not mix together but form as two layers in the beaker.

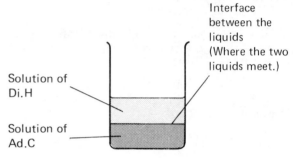

▲ **What to write**

5. Say where you think the nylon will form:

   i. in the water layer;

   ii. in the tetrachloromethane layer;

   iii. at the interface between the two liquids.

6. What do you think you will see happen where the two liquids meet?

Unit 4
Making Nylon — an experiment

■ What to do

Note: The materials for this experiment are expensive, and so take care not to waste them.

Collect:
(a) a 10 cm³ beaker
a special wire
a bottle of Di.H solution
and a bottle of Ad.C solution.

(b) First put 2 pipettes full of Ad.C solution into the beaker.

(c) Very carefully add 2 pipettes full of Di.H solution. *Do not mix them!*

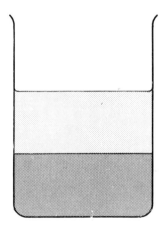

(d) The nylon forms at the interface between the two liquids. With the special wire pull out the nylon from the interface.

★ Information

The Ad.C and Di.H solutions are liquids. The molecules in them are free to move and tumble over each other. When they are brought together, however, they form the solid nylon.

▲ What to write

1. What has happened to the molecules of Di.H and Ad.C that changes them from being liquids to a solid?

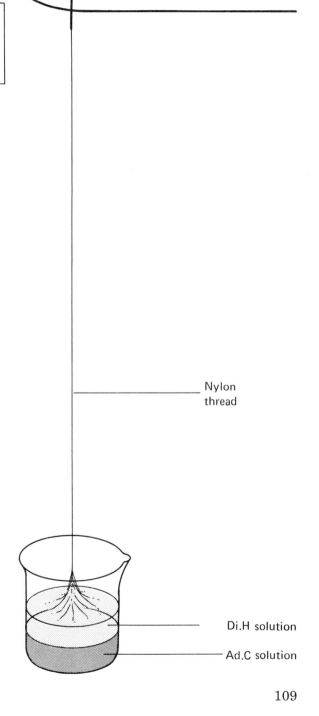

Nylon thread

Di.H solution

Ad.C solution

Unit 5
Poly compounds

★ **Information**

One of the important developments of the last 20 years has been the increasing use of man-made polymers and plastics. Here are the names of some of them.

Common name	Chemical name
Polythene	polyethylene
P.V.C.	polyvinylchloride
Perspex	polymethyl methacrylate
Nylon	polyamide
Terylene	polyacetate

You will notice that the chemical name for these compounds begins with the word *poly-*. This is a Greek word meaning *many*.
The name polyvinyl chloride means "many vinyl chlorides" because it is made by joining many vinyl chlorides together.

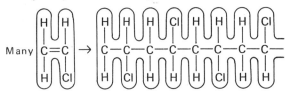

▲ **What to write**

1. What exactly does the word polystyrene mean?

★ **Information**

The process in which many molecules of the same type are joined together to make a very big molecule is called *polymerisation*.

Such very big molecules are all given the general name *polymers*.

The small molecules that join together to make the polymers are called *monomers*.

Different kinds of monomers make different kinds of polymers.
Many molecules of the monomer styrene:

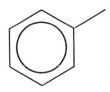

join together to make the polymer polystyrene.

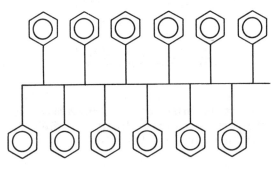

▲ **What to write**

2. Polythene is formed by many molecules of a gas called ethene joining together. Which is the monomer and which is the polymer in this example?

3. What is the name of the monomer that forms polystyrene?

4. What is the name of the monomer that forms polyvinyl chloride?

Unit 6
Breaking down and repolymerising Perspex

★ **Information**

Perspex is a polymer. It is made up of hundreds of molecules of the monomer methyl methacrylate MM joined together.

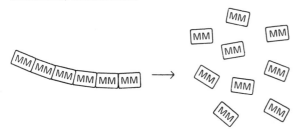

In this experiment you will be breaking up the polymer perspex into the individual monomer molecules.

IMPORTANT: THIS EXPERIMENT MUST ONLY BE DONE IN A "FUME CUPBOARD".

■ **What to do**

(a) Collect:
two test tubes
a delivery tube
some Perspex chips
a beaker of ice and water.

(b) Put the apparatus together as in the diagram.

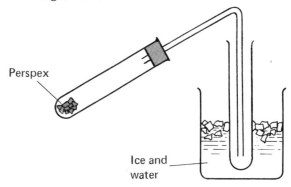

(c) Heat the Perspex carefully. Do not heat it too much or it will go black.

WARNING: DO NOT BREATHE THE FUMES AS THEY ARE POISONOUS.

▲ **What to write**

1. Draw a diagram of the apparatus and mark on it where the polymer is and where the monomer is.
2. Describe what the Perspex monomer looks like.
3. What is the purpose of the ice?
4. Explain why the polymer is a solid and the monomer is a liquid.

Putting the Perspex molecules back together again.

★ **Information**

If the Perspex monomer molecules were left for many years they would eventually join together again to make Perspex. If we use a catalyst and a higher temperature, however, we can make the molecules join together again much more quickly.

■ **What to do**

(d) Divide your sample of Perspex monomer into two test tubes.

(e) To one of the tubes add one small spatula measure of "Perspex catalyst".

(f) Plug both tubes with cotton wool to discourage the harmful vapour from escaping.

(g) Mark both tubes with your initials.

(h) Take the tubes to your teacher so that they can be put into an oven set at 90°C for about an hour.

(i) Collect your tubes in your next lesson.

▲ **What to write**

5. If we want the Perspex monomer molecules to join together again more quickly which two things can we do?

6. When your test tubes are returned to you in your next lesson, what will you expect to see in the two tubes?

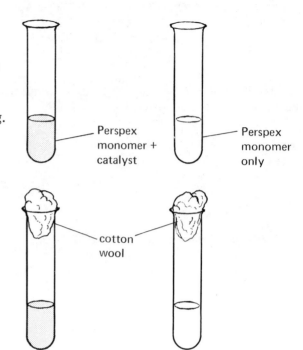

Unit 7
The polymer starch

★ **Information**

The chemical *starch* is a polymer of the monomer *glucose*.

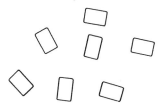

molecules of glucose

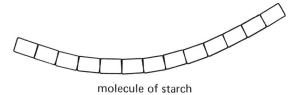

molecule of starch

■ **What to do**

(a) Collect a test tube, and add 1 cm depth of fresh starch suspension.

(b) Add 2 or 3 drops of iodine solution.

(c) This is the test for starch. If the iodine goes dark blue, it means starch is present.

▲ **What to write**

1. What is the name of the polymer you are using in this experiment?
2. What is the name of its monomer?
3. What happens to iodine if starch is present?

■ **What to do**

(d) Some of the starch may have changed into its monomer, glucose. We can test for glucose by heating with Benedict's Solution.

(e) Wash out your test tube, and put another 1 cm depth of fresh starch suspension into it.

(f) To see if any starch has changed to glucose, add 1 cm depth of Benedict's Solution.

(g) Now gently boil the mixture.
DON'T POINT THE TEST TUBE AT ANYBODY.
If the liquid stays blue, none of the starch has changed into glucose molecules.

★ **Information**

Heating a substance with Benedict's Solution is the test for glucose.

Final colour in test tube		
Blue	Green	Yellow/red
No glucose	A little glucose	A lot of glucose

▲ **What to write**

4. Did you detect any glucose molecules in the starch suspension?
5. How can you tell whether there is a lot, or whether there is a little glucose present?

■ **What to do**

(h) Now test the "One Month Old Starch Suspension" for starch and for glucose.

▲ **What to write**

6. Was there any glucose in the old starch suspension?
7. Was there still some starch left, even after 1 month?
8. Does it seem to be true that starch molecules are slowly breaking up into glucose molecules?

Unit 8
A catalyst again

★ **Information**

You have found that the polymer starch slowly breaks up into the monomer glucose.

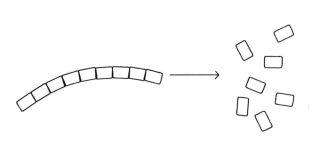

We'll now try to speed this up with a catalyst.

■ **What to do**

(a) Put about 2 cm depth of fresh starch suspension into a clean test tube.

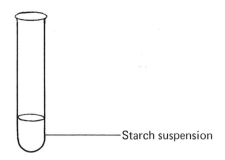

(b) Add about 2 cm depth of Amylase Solution.

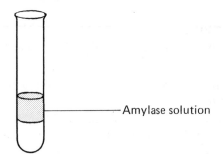

(c) Now add 5 drops of iodine solution.

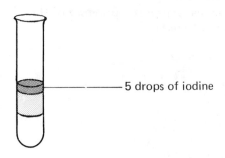

(d) Leave the test tube in a beaker of warm water for 10 minutes.

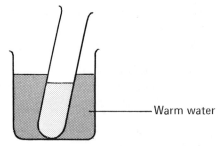

(e) Watch the colour in the test tube from time to time.

(f) While you wait, answer questions 1 to 4.

▲ **What to write**

1. What is the chemical reaction you are trying to speed up?
2. Which two methods are you using to try to speed up the reaction?
3. What is the name of the catalyst used in this experiment?
4. How can you tell when all the starch has been broken up?

■ **What to do**

(g) When all the starch has broken up, test for glucose by heating with Benedict's Solution.

▲ **What to write**

5. Did the Amylase solution speed up the reaction?

Extension Work 1
Proteins

★ Information

Protein molecules are polymers, their monomers being *amino acids*.

You are given a suspension of insoluble protein. If the protein is changed into amino acid, then the suspension goes clear, because the amino acid molecules are soluble.

▲ What to write

1. What polymer are you using in this experiment?

2. What is its monomer?

3. How can you tell whether the polymer has been changed into its monomer?

(a) Put about 3 cm depth of protein suspension into a test tube.

(b) Add 5 drops of dilute hydrochloric acid.

(c) Now add 2 cm depth of the catalyst 'Pepsin solution'.

(d) Put the test tube into a beaker of warm water, and watch what happens.

▲ What to write

4. What did you see happening in the test tube?

5. What was happening to the polymer molecules?

6. How would you show that the catalyst did, in fact, speed up the reaction, and that the same changes, at the same rate, would not have happened anyway?

★ Information

In previous work you have seen that all substances of the "living world" consist partly of carbon atoms. Many of these substances are used by man either as fuels or to make plastics or other polymers. Man's chief sources of these materials are coal and petroleum oil. The next two experiments look at these two substances and the way that they can be separated into different substances for different uses.

Extension Work 2
Coal

■ What to do

(a) Collect the pieces of apparatus shown in the diagram.

(b) Put 2 or 3 pieces of coal in the test tube.

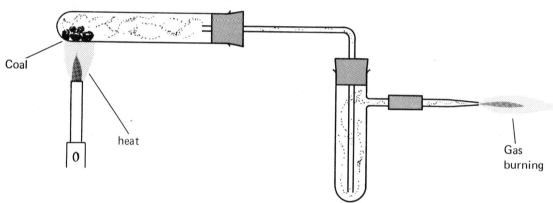

(c) Heat the coal. See if you can get the coal gas to burn. Look out for the other substances formed.

There will be: one substance left in the test tube.
two substances in the side arm tube.
and the gas you burn.

▲ What to write

1. Draw a diagram of the apparatus you used.

2. Describe by labels on your diagram the four substances that coal is broken up into.

3. By using the various books available find out about some of the uses of the materials obtained from coal.

Extension Work 3
Oil

★ **Information**

Petroleum oil is in fact a mixture of many similar molecules but of varying sizes and shapes. Here are a couple of examples.

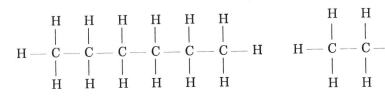
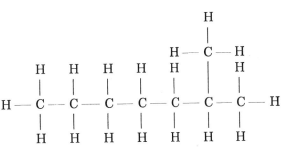

In an oil refinery the mixture is separated into different fractions that have similar boiling points. This process is called fractional distillation.

■ **What to do**

The experiment is rather difficult to do and you may find that you need to repeat it before you obtain good results.

(a) Collect and set up the apparatus illustrated below and also have ready 4 clean and *dry* test tubes in a test tube rack.

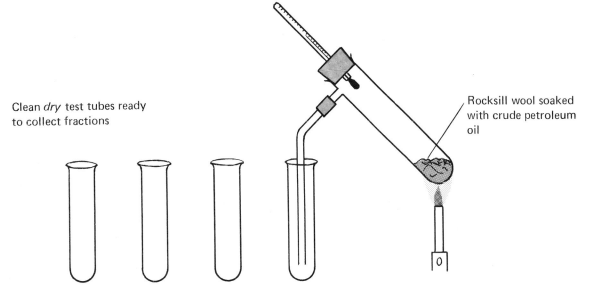

Clean *dry* test tubes ready to collect fractions

Rocksill wool soaked with crude petroleum oil

(b) Start by heating the tube of crude oil very gently so that only that portion of the oil that boils at a low temperature boils off first. Keep an eye on the thermometer and collect the fractions distilling between the temperatures shown in Table 10.
Note: It is quite all right to stop heating for a while as you change the test tubes over.

(c) Once you have collected your four fractions and the residue left in the heating tube, carry out tests on them so that you can fill in Table 10.

IMPORTANT: ASK YOUR TEACHER HOW TO *SAFELY* TEST FOR "EASE OF IGNITION".

Table 10

Fraction	Distilling between temperatures	Colour	Viscosity	Ease of ignition
1	Room temp. – 70°C			
2	70°C - 120°C			
3	120°C - 170°C			
4	170°C - 220°C			
Residue	over 220°C			

★ **Information**

Viscosity refers to the thickness of the liquid or in other words how "runny" it is. Treacle is said to have a high viscosity and water has a low viscosity.

Ease of ignition refers to the readiness with which the liquid burns.

▲ **What to write**

1. Draw a diagram of the apparatus you used.

2. What is the function of the thermometer?

3. What property of the different molecules in the oil were you using to separate the fractions?

4. Copy Table 10 and complete it with your results.

5. Use the various books and pamphlets provided to find out what the different fractions of crude oil are used for.

6. Which fraction of oil do you think had the biggest molecules?

15

Grub, Guts and Spit

Unit 1
Food tests

▲ What to write

1. How do you test for starch?
2. How do you test for glucose?

If you have trouble answering these questions, look back to Chapter 14, Units 7 and 8.

■ What to do

(a) Put a few bread crumbs into a test tube.
(b) Add a few drops of iodine solution.
(c) Test some of the other foods provided for starch.

▲ What to write

3. Copy out Table 11.

Table 11

Food	Starch	Glucose

Put a tick (√) for starch present.
Put a cross (✕) for starch absent.

4. Fill in your results, to show which substances did contain starch, and which did not.

■ What to do

(d) Carry out a test for glucose on the lemonade provided, by heating some with Benedict's solution.

(e) To test solids for glucose, put a little of the solid into a test tube, shake it with a little water, and then add Benedict's Solution, as before.

Now test some of the other foods for glucose.

▲ What to write

5. Complete Table 11.

Unit 2
The model body

★ **Information**

You are going to see what happens to food inside your gut (alimentary canal).

Your alimentary canal is a long tube running through your body, and its main parts are the mouth, the stomach, the intestines, and the hole at the end, called the anus.

Food that you take in through your mouth will pass out through the anus, unless the body changes the food.

The solid waste, that the body cannot change, and which we get rid of down the lavatory, is called the faeces.

▲ **What to write**

1. What is the proper biological word for gut?

2. Put the following parts of your alimentary canal (gut) in the right order, starting at the mouth: *stomach, anus, mouth, intestines.*

3. Food is changed while in the alimentary canal, but some of it passes out through the anus. What is this waste called?

4. You are going, from time to time, to be using a model body consisting of:
    a boiling tube ('skin')
    warm water ('blood and muscles')
    visking tubing ('imitation intestines')

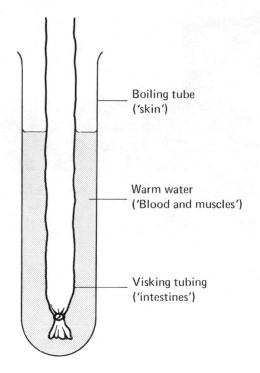

Why should you use *warm* water?

5. To be most realistic, at which of the following temperatures ought the water in the boiling tube to be?

    0°C, 15°C, 37°C, 59°C, 100°C?

    Give a reason for your answer.

Unit 3
Imitation food

★ **Information**

You are going to make your own 'model body' using a boiling tube, warm water, and visking tubing.

■ **What to do**

(a) Fetch a boiling tube — this represents your skin.

(b) ¾ fill it with warm water — this represents your blood.
Put the boiling tube into the test tube rack.

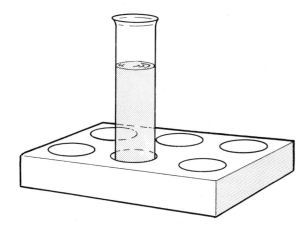

(c) Now fetch a piece of visking tubing. This represents your intestines.

(d) Carefully put a little 'Imitation food A' into the visking tubing.

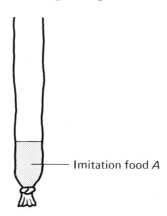

(e) Now add a little 'Imitation Food B'.

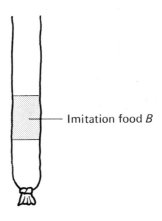

(f) Fasten the top of the visking tube with a paper clip.

(g) Wash the outside of the visking tube under the tap.

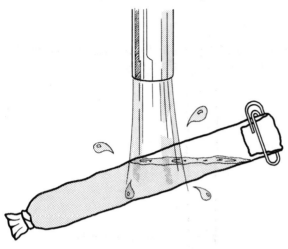

(h) Put the 'intestines' with its 'food' into the 'body' (boiling tube of water).

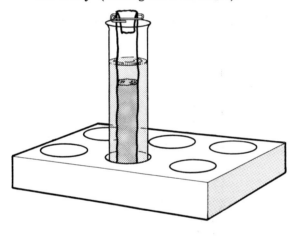

(i) During the next 10 minutes, make a large labelled diagram (about ½ page) of your apparatus.

Label the 'skin', 'blood' and 'intestines'.

Keep your eye on the experiment, to see if anything is happening.

▲ What to write

1. What did you see happening?

2. Did any of the red 'food' (food A) pass through the side wall of the 'intestines', into the 'blood'?

3. Did any of the blue 'food' (food B) pass through this wall?

4. Did *all* the food pass through this wall?

5. In a real body, what would eventually happen to the food which does not pass through the side wall of the intestines?

6. For each of the following statements, say whether it is *true* or *untrue*.

   i. All food taken into the alimentary canal passes through the side wall of the intestines.

   ii. All the food taken into the alimentary canal passes through the anus.

   iii. Some of the food taken into the alimentary canal passes through the side wall of the intestines, and some passes through the anus in the form of faeces.

   iv. The food which passes through the intestines wall is then moved around the body in the blood.

Unit 4
Real food

★ **Information**

You have been working with 'imitation food' so far.

You have discovered, in previous work, that starch and glucose (sugar) are present in many of the foods you eat.

Let's try feeding starch and glucose (sugar) to the visking tubing.

■ **What to do**

(a) Set up your experiment, as in Unit 3, but instead of using imitation food, use the mixture of starch and sugar (glucose) solution.

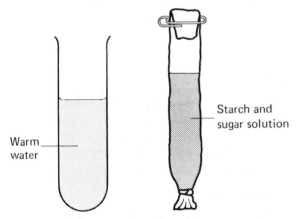

Don't forget to wash the outside of the visking tubing very carefully before putting it into the boiling tube.

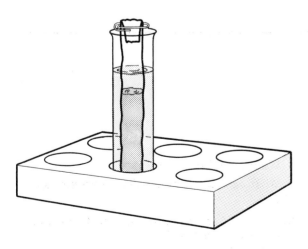

(b) You are going to find out whether any of the starch or sugar has passed into the blood (water in the boiling tube).

But you will again have to wait for 10 minutes.

▲ **What to write**

1. In the previous unit, you found that some of the imitation food would go through the visking tubing, and some would not.

   What is the point of the experiment in this unit (Unit 4)?

2. Why are you told to wait for 10 minutes?

■ **What to do**

(c) After 10 minutes, take the visking tube out of the boiling tube, and put it into the container marked 'Dirty visking tubing'.

(d) Pour 2 cm depth of the water from the boiling tube into each of two test tubes.

(e) Test the water in one of the test tubes with a drop of iodine solution, to see if any starch is present.

(f) Test the water in the other test tube for sugar, by heating with Benedict's Solution.

Remember the safety precautions.

▲ **What to write**

3. Was starch present in the water?

4. Was there any sugar present?

5. You may find that your results are different from other people's. If they are, how are you going to decide who is right?

6. Is it true that the visking tubing allowed one food substance to pass through, and not the other food substance?

7. Which molecules, those of starch, or those of sugar, could get through the visking tubing?

8. Why do you think the sugar molecules could get through the wall of the visking tubing, whereas the starch molecules could not?

Unit 5
Useless food?

★ **Information**

You have discovered that starch molecules *cannot* go through the visking tubing, but that sugar molecules *can*.

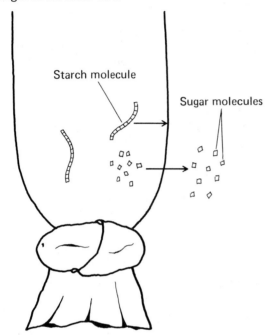

■ **What to do**

(a) Imagine that you have just invented a microscope powerful enough to see sugar molecules.

(b) Imagine that you are now looking down the microscope at some visking tubing filled with a mixture of starch and sugar solutions.

▲ **What to write**

1. Make a drawing of a little part of the visking tubing, as seen down your new microscope, showing how you think sugar molecules manage to get through the visking tubing, whereas starch molecules cannot.

   Remember:
   i. what sugar molecules are like
   ii. how big starch molecules are.

★ **Information**

Starch cannot get into the blood system, because its molecules are too big to get through the side wall of the intestines.

Unless our body can do something about it, all the starch in our food will go along the gut, and will be got rid of in the faeces.
This part of our food would be wasted.

▲ **What to write**

2. Suggest what the body might try to do to the starch molecules, to make them useful to the body. (Draw a diagram, if you like.)

Unit 6
What do we do with starch?

★ **Information**

In previous work (in chapter 14), you have found that starch can be broken up into smaller molecules.

▲ **What to write**

1. What can starch molecules be broken up into?

2. Did you find that this was a fast reaction or a slow reaction?

3. What is the name of the catalyst you used to speed up the reaction?

4. Do you think it would be useful for your body to have this catalyst in the alimentary canal? Give reasons for your answer.

★ **Information**

The substance Amylase helps to change starch molecules into sugar molecules.

Amylase is an *enzyme*, which means a biological catalyst.

Let's see if you can detect Amylase in your saliva (spit).

■ **What to do**

(a) Fetch a 100 cm³ beaker.

(b) Dribble a little saliva into the beaker.

(c) Set up a test tube as follows:

　i. Put about 1 cm depth of starch solution into the test tube.

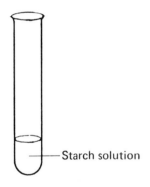

　ii. Add about 5 drops of iodine solution.

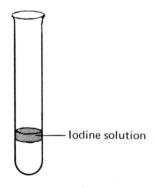

　iii. Add about 2 cm depth of saliva. (Don't measure the bubbles.)

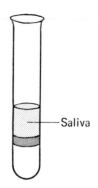

　iv. Stir with a glass rod.

　v. Put the test tube into a beaker of warm water.

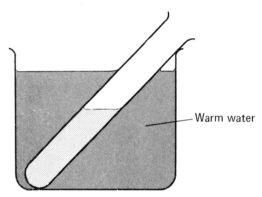

　vi. Leave the test tube set up, and see what happens to the colour.

▲ **What to write**

5. What happened to the colour in the test tube?

6. How do you know that there was starch in the test tube at the beginning of the experiment?

7. Was there any starch at the end of the experiment?

8. What happened to the starch molecules, do you think?

9. How could you test whether the starch molecules have been changed into sugar molecules? Try it.

10. Does saliva contain Amylase, or, at any rate, something like it? (Answer *Yes* or *No*.)

Extension Work 1
Digest your eggs

★ **Information**

Proteins, too, have to be broken up in the alimentary canal.

'Egg White Suspension' contains a protein. When the protein is digested, the mixture goes clear.

■ **What to do**

(a) Put about 3 cm depth of Egg White Suspension into a test tube.

(b) Add *5 drops* of dilute hydrochloric acid.

(c) Now add 2 cm depth of the enzyme (catalyst) *pepsin*, which is present in our stomachs.

(d) Put the test tube into a beaker of warm water.

(e) Watch to see if digestion takes place. (i.e. see if the mixture goes clear.)

▲ **What to write**

1. Describe:
   i. What you did.
   ii. What the point of the experiment was.
   iii. What your results were.

Further ideas

(a) Does pepsin work without acid?

(b) Does amylase work on starch if you add 5 drops of acid?

(c) Will amylase digest the protein in Egg White Suspension?

Extension Work 2
Acid guts

★ **Information**

You have found that pepsin digests proteins, as long as you add about 5 drops of acid.

Is the exact amount of acid important, or doesn't it matter too much?

The *degree of acidity* of a mixture is measured by its *pH*.

pH 1 – 7 means the mixture is acid.

pH 7 means the mixture is neutral.

pH 7 – 14 means the mixture is alkaline.

You can find out the pH of a mixture by using indicator paper, which turns a different colour for different pH.

▲ **What to write**

1. Explain what you are trying to find out.

2. Work out how you are going to find out the answer.
   (Once you have worked it out carefully, discuss your method with your teacher.)

3. At what pH did you find pepsin worked best?

4. How would you find out at what pH amylase works best? Try it if you like.

5. At what pH did amylase work best?

6. Amylase works in the mouth. At what pH would you expect the mouth to be?

7. Pepsin works in the stomach. At what pH would you expect the stomach to be?

8. Stomach powders for indigestion are alkalis. What, then, would you think is the cause of indigestion?

Extension Work 3
Indigestion

★ **Information**

When someone suffers from 'indigestion', it means that his food won't digest properly in his stomach, because there is too much acid in the stomach for the pepsin to work.

To overcome this, he takes 'Stomach Powder', which is an alkali, and therefore neutralizes (gets rid of) the extra acid.

Imagine that you are the Managing Director of a firm making a brand of stomach powder.

■ **What to do**

(a) Invent a good name for your stomach powder.

(b) Make an advert which is to appear in a Sunday newspaper colour supplement.

(c) Write down some outline ideas for a commercial for your stomach powder, which is to appear on television.

Extension Work 4
Saliva versus starch

★ **Information**

You have found that amylase in your saliva digests starch.

See if you can get saliva to work inside your visking tubing, so that sugar appears in the 'blood' (water in the boiling tube).

■ **What to do**

(a) Put a mixture of starch and saliva into a visking tube.

(b) Close the visking tube, and wash it.

(c) Put the visking tube into a boiling tube of warm water.

(d) Keep the boiling tube warm, by putting it into a beaker of warm water.

(e) After 20 minutes or so, test the 'blood' for sugar.

▲ **What to write**

1. Make a neat and large diagram of your apparatus.

2. What results did you get?

Equipment List

1 Let there be light ...

Part A

Unit 1
Light source
Small screen and stand
Low voltage supply or power-pack
Small piece of card
Pair of scissors

Unit 2
Light source
Low voltage supply or power-pack
Plane mirror and support
Large sheet of white paper

Unit 3 (Demonstration)
Demonstration smoke-box
Demonstration glass lens for smoke box
Compact light source
Cardboard to make smoke
Smoke "puffer"

Unit 4
Aluminium saucepan or similar container
Coin
Small piece of Plasticine
Pencil

Parts B and C

Unit 5
+7D spherical lens
Small screen and stand
Large sheet of black paper
½-metre ruler

Unit 6
+7D spherical lens

Unit 7
½-metre ruler
Sheet of greaseproof paper
Some Plasticine
2 +7D lenses
1 +14D lenses

Unit 8
Small 60° prism
Light source
Low voltage supply or power-pack
Small screen and stand
Large sheet of white paper
Small sheet of white paper
Large sheet of black paper

Unit 9
"Colour mixer"

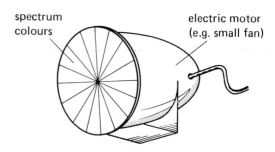

"Colour mixer"

2 ... And there was light

Much of the following is needed for every Unit in this section.

Two light sources
Low voltage supply or power-pack
Card with one thin slit
Card without a slit
Card with many slits ("comb")
Card with three thin slits
A small glass 60° prism
Large sheet of black paper
Large sheet of white paper
Two +7D cylindrical lenses
+17D cylindrical lens
—17D cylindrical lens
Plane mirror and support

3 Camera and eye

Unit 1
Pinhole camera box
Small sheet of black paper
Large sheet of black paper
Small sheet of greaseproof paper
A pin
Adhesive tape
Pair of scissors

Unit 2
As Unit 1

Unit 3
As Unit 1

Unit 4
As Unit 1 plus a +7D spherical lens

Unit 5
1 eye
1 scalpel
1 pair of forceps
1 dish for dissecting in

Unit 6
1 bright lamp (which can be switched on and off) should be available

Extension Work 1
None

Extension Work 2
None

4 Forces

Unit 1
None

Unit 2

Exhibit 1 (2 sets per class)
Long ball-ended magnet
A cork large enough to make the ball-ended magnet float
Plastic tank
Bar magnet
Plasticine to hold the bar magnet on the side of the plastic tank

Exhibit 2 (3 sets per class)
Cellulose acetate strip
Polythene strip
Cradle for supporting the two previous items
½-metre length of cotton or nylon fishing line
Duster
Retort stand and clamp

Exhibit 3 (2 sets per class)
Measuring cylinder filled with glycerol
Jar of small nails
Wooden board with sandpaper and smooth cartridge paper surfaces

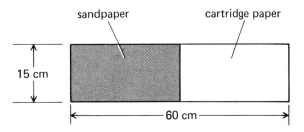

Wooden block with bent nail or hook

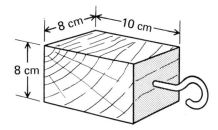

Hand lens or binocular microscope

Exhibit 4
Turntable *(1 per class)*
1 kg mass *(2 per class)*

Exhibit 5 (4 per class)
Luggage strap
Forcemeter (0–15 newton)

Exhibit 6 (4 per class)
Small cardboard box, labelled *A* and sealed with string and sellotape. The box contains slotted masses to give it a total mass of 400 g.

Cardboard box similar to the previous item, of total mass 800 g, labelled B.
Forcemeter (0 - 10 newton or 0 - 15 newton)
100 g slotted mass (20 per class)
Holder for slotted masses
A weak spring which has a large extension when used to suspend a 1 kg mass

Unit 3

Dynamics trolley with post
Elastic cord with rings tied to the ends

Unit 4

Postcard
Metal washer
Plastic beaker
Wooden block with nail

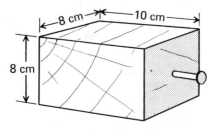

Wooden block with nail

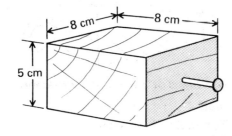

Retort stand and clamp
G-clamp
80 cm length of thread
Elastic cord with rings tied to the ends

Extension Work 1 and 2

Wooden block with nail similar to Unit 4, item 4
Clamp from a retort stand
Plasticine available

Sheet of cardboard

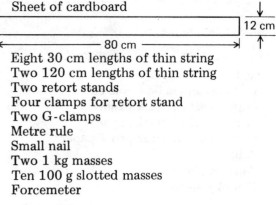

Eight 30 cm lengths of thin string
Two 120 cm lengths of thin string
Two retort stands
Four clamps for retort stand
Two G-clamps
Metre rule
Small nail
Two 1 kg masses
Ten 100 g slotted masses
Forcemeter

Extension Work 3

No apparatus needed

5 To the moon

Unit 1
No apparatus needed

Unit 2
No apparatus needed

Unit 3
A cardboard box, cut, glued and painted to look like a 1 kg mass. Material is inserted to make its total mass about 170 g.
A tray containing sand and pebbles (the moon's surface) and the model 1 kg mass. The whole thing labelled "Model of a 1 kg block on the Moon's surface".
Forcemeter (0 - 10 newtons)

Unit 4
No apparatus needed

Extension Work 1
No apparatus needed

6 Making things work

Unit 1
No apparatus needed

Unit 2
No apparatus needed

Unit 3
  1 kg mass
  Wooden block with nail as for Forces
  Unit 4 (10 × 8 cm × 8 cm)
  G-clamp
  Forcemeter (0 – 10 newton or 0 – 15 newton)
  150 cm length of string
  Metre rule
  Stop clock

Extension Work 1
  Small electric motor (6 volt)
  Low voltage supply
  Two leads
  1 metre length of thread
  Four 50 g slotted masses
  Holder for slotted mass
  G-clamp

7 'Movement'

Unit 1
  Dynamics trolley
  Runway
  Stop-clock
  Metre rule
  Roll of Sellotape

Unit 2
  Polystyrene or table tennis ball
  Metal or wooden block

Unit 3
  Ticker-tape vibrator
  Low voltage supply (a.c.)
  Roll of ticker-tape (adhesive)
  Two leads
  Runway
  Dynamics trolley
  Metre rule
  Large sheet of sugar paper

Extension Work 1
  Dynamics trolley and post
  Low voltage supply
  Ticker-tape vibrator
  Two leads
  Roll of ticker-tape
  Two elastic cords with rings attached to the ends
  Eight 4 cm squares of hardboard
  Large sheet of sugar paper

Extension Work 2
  Two CO_2 pucks
  Table with smooth surface and raised edges
  CO_2 cylinder with dry ice attachment

8 Breaking up is hard to do

Unit 1
  1 tripod
  1 pipe clay triangle
  1 asbestos mat
  1 bunsen burner
  5 or 6 pieces of asbestos paper
  Access to samples of wood, sand (preferably preheated so that it does not change colour when the students heat it), flour, grass, salt, sugar.

Unit 2
  1 test tube
  Wood splints
  Cobalt chloride paper (blue)
  Dropping pipette
  A small test tube for use in the lime water test
  Spatula
  Access to zinc nitrate

Unit 3
  Several test tubes
  All other items as for Unit 2

  Access to the following substances:
  copper(II) carbonate
  copper(II) oxide
  copper(II) nitrate
  zinc carbonate
  iron carbonate
  iron(III) nitrate
  magnesium oxide

Unit 4
  1 combustion tube (A test tube with small hole blown in the end)
  1 rubber connecting tube with cork attached to fit the combustion tube
  1 bunsen burner
  1 clamp and clamp stand
  1 asbestos mat

  Access to a sample of copper(II) oxide

Extension Work 1
2 test tubes and a delivery tube

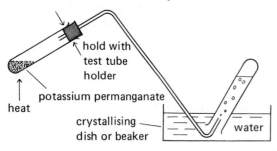

Test tube holder
Wooden splints
Indicator paper
Potassium permanganate(VII)

Extension Work 2
As for Unit 1 plus copper carbonate and lime water.

9 Atoms

Some Plasticine

10 Elementary my dear compound

Unit 1
1 pair of tongs
A bunsen burner
Asbestos mat
1 or 2 pieces of magnesium ribbon

Unit 2
1 test tube rack
2 test tubes
Access to bottles of dilute hydrochloric acid
1 or 2 pieces of magnesium ribbon
Access to apparatus for testing the electrical conductivity of solid materials

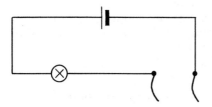

Unit 3
1 combustion spoon, "Nuffield type"
A bunsen burner
Asbestos mat
1 boiling tube full of oxygen + bung
Access to powdered carbon

Access to lime water.

Extension Work 1
Access to samples of the following for
    flame testing.   copper(II) chloride
    sodium chloride   strontium chloride
    calcium chloride   barium chloride
Also a sample marked unknown but containing one of the metal ions listed above.
1 bunsen burner
A beaker
Access to distilled water
A sharp pencil

Extension Work 2
1 glass rod
1 glass plate (approx. 10 cm by 15 cm is ideal)
Access to dropping bottles of:
    sodium hydroxide
    copper(II) sulphate
    iron(II) sulphate
    iron(III) sulphate
    zinc sulphate
    magnesium sulphate
Also a dropping bottle containing a solution containing one of the metal ions listed above, labelled Z.

Extension Work 3
1 test tube rack
6 test tubes
1 spatula
1 large test tube and bung to fit
1 combustion spoon
1 straw
1 bunsen burner
Access to a diluted solution of universal indicator
Samples of
    copper(II) oxide
    calcium oxide
    magnesium oxide
    iron(III) oxide
    phosphorus oxide (The teacher may prefer to give this out)
    sulphur

11 Reactivity

Unit 1. Demonstration
Glass beakers or troughs
Samples of the following metals:
    sodium
    magnesium
    copper
    potassium
    iron
    lithium
    zinc.

Unit 2
4 test tubes
A test tube rack
Wooden splints
Access to the following
    copper turnings
    zinc pieces
    magnesium ribbon
    iron filings (coarse)
    2M hydrochloric acid

Unit 3
1 evaporating basin
1 test tube
1 piece of magnesium ribbon about 5 cm long
2M hydrochloric acid
Bunsen burner
Tripod
Gauze
Asbestos mat

Unit 4
1 test tube
1 spatula
Access to 2M hydrochloric acid
copper(II) oxide

Unit 5
4 petri dishes
Access to
    pieces of magnesium ribbon
    shiny iron nails
    pieces of copper
    pieces of zinc
solutions of
    magnesium chloride
    copper (II) chloride
    zinc chloride
    iron (III) chloride

Extension Work 1
1 blow pipe
1 carbon block
Bunsen burner
Asbestos mat
Access to lead oxide
  and a means of making a cavity in the carbon block

Extension Work 2
Charts on iron and steel making
Film loop on the blast furnace is also useful

Extension Work 3
None

12 Rates of reaction

Unit 1
1 spatula
Two 100 cm^3 beakers
Access to cold and hot water
Access to either commercial health salts
  or a prepared mixture of citric acid and sodium hydrogencarbonate

(*Note*. The prepared mixture of citric acid and sodium hydrogencarbonate will not keep for longer than one day.)

Unit 2
Access to the following:
    cold dilute hydrochloric acid
    warm dilute hydrochloric acid
    pieces of zinc foil cut up
    test tubes

Unit 3
1 test tube rack
2 test tubes
1 spatula
Access to pieces of copper
Access to zinc shavings
Access to dilute hydrochloric acid

Extension Work 1
1 test tube
Two 100 cm^3 beakers
2 spatulas

Access to dilute hydrochloric acid and calcium carbonate

Extension Work 2
  Two 100 cm³ conical flasks
  1 test tube rack
  2 test tubes
  Access to dilute hydrochloric acid and marble chips sorted into large and small sizes
  A means of weighing out 10 g portions of marble chips

13 Get a move on

Unit 1
No apparatus is required but wallcharts or film loops on the movement of particles in solid/liquid/gas could be useful.

Unit 2
  Graph paper
  1 tripod
  1 gauze
  1 bunsen burner
  1 asbestos mat
  1 clamp and clamp stand
  1 250 cm³ beaker
  1 0 - 100°C thermometer
  1 test tube containing about 3 cm depth of naphthalene
  Stop clock

Unit 3
If the students do not use exercise books some alternative will need to be found for the "flick book".

Unit 4
No apparatus required but questions from students may need access to a book of data for answering.

Unit 5
No apparatus is essential but any visual displays of atomic structures would be useful.

Unit 6
  Bunsen burner
  Asbestos mat
  Test tube holder
Test tubes containing substances for the students to identify as giant or molecular structures. Suitable substances would be:

  candle wax   magnesium oxide
  acetamide   copper oxide
  carbon graphite   iodine

Extension Work 1
Apparatus for electrolysing copper chloride solution using carbon rods.

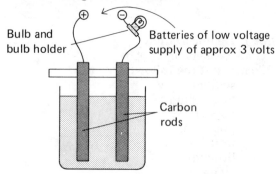

Also: copper chloride, salt, sugar and any other suitable soluble substance for electrolysing.

Extension Work 2
Set up a display of metals showing the crystal structure of metals e.g. mercury in silver nitrate leads to the formation of silver crystals, copper foil in silver nitrate solution and zinc in lead nitrate solution

Extension Work 3
Apparatus and materials for the students to grow their own crystals.

14 Merging molecules

Unit 1
Samples of sulphur for the students to look at, preferably crystallized rhombic sulphur.

Unit 2
  Test tube
  Test tube holder
  Bunsen burner
  250 cm³ beaker
  Access to crushed roll sulphur
  Asbestos mat
  Cloth, kept damp

Unit 3
Demonstration kit for showing the polymerisation of nylon. Consists of plastic tray, or other container with a smooth surface; small blocks of balsa wood with small 1 cm magnets stuck to them with an epoxy adhesive.
  Approximately 30 blocks should be sufficient for each kit. Half of the blocks have the

side with the S poles uppermost marked "Ad.C" and the other half have the side with the N poles uppermost marked "Di.H".

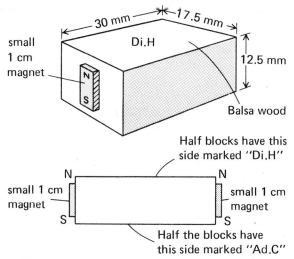

Then with the marked faces uppermost, gentle shaking of the tray from side to side will cause the blocks to move about and attach themselves to each other in the correct order.

| Ad.C | Di.H | Ad.C | Di.H | Ad.C | Di.H | ... |

Unit 4
One 10 cm³ beaker
One piece of bent wire, to pull out the thread of nylon

Access to 5% solution of adipyl chloride in tetrachloromethane, and to 5% solution of 1-6 diaminohexane in water

These should be dispensed by means of dropping pipettes (or bottles with dropping pipettes attached) in order to conserve these expensive chemicals.

Unit 5
No apparatus required

Unit 6
2 test tubes
A delivery tube
One 250 cm³ beaker

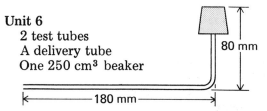

Access to chips of Perspex, small enough to fit in the test tube
Ice

Repolymerisation
1 extra test tube
Access to cotton wool
Perspex catalyst (lauroyl peroxide)

The teacher will require a fume cupboard and a means of keeping the Perspex hot (approx. 90°C) for about 1 hour. This can be done by means of an oven set in the fume cupboard or by standing the tubes in very hot water.

Unit 7
1 test tube
1 test tube holder
1% starch suspension, marked 'Fresh Starch'.
1 month old starch suspension, or starch suspension contaminated with glucose, marked '1 Month Old Starch'.
Iodine in potassium iodide solution in dropping pipette reagent bottle
Benedict's solution
Bunsen and asbestos square

Unit 8
1 test tube
1 test tube holder
1% starch suspension
1% Amylase solution
Iodine in potassium iodide solution
1 beaker
Benedict's solution
Bunsen, and asbestos square

Access to warm water

Extension Work 1
Egg white suspension —
(thoroughly stir the egg white of 1 egg into 500 cm³ of water. Boil and cool. Marked 'Protein Suspension'.)
1 test tube
Dilute hydrochloric acid
1% pepsin solution
1 beaker

Access to warm water

Extension Work 2
Small pieces of coal that will fit in a test tube
Bunsen burner

Clamps and stands
A test tube, side arm tube, and delivery tubes to make the apparatus illustrated.

Extension Work 3
Side arm test tube
Rocksill wool soaked with crude petroleum oil
0 – 250°C thermometer
4 clean dry test tubes
Bunsen burner

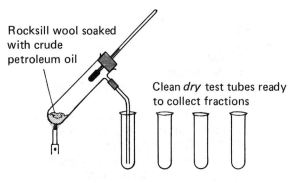

Evaporating dish in which the oil fractions are burned

15 Grub, guts and spit

Unit 1
Bread crumbs
Crushed cereal
Crushed biscuits, contaminated with glucose
'Lemonade' — glucose solution
Iodine in potassium iodide solution
Benedict's solution
Bunsen and asbestos mat
1 test tube
1 test tube holder

Unit 2
None

Unit 3
1 boiling tube
1 test tube rack, capable of holding a boiling tube
1 piece of visking tubing, 14 mm diameter, about 20 cm long, with a knot in one end.
'Imitation Food A' — potassium permanganate solution
'Imitation Food B' — methylene blue solution
Paper clip

Unit 4
1 boiling tube
1 test tube rack, capable of holding a boiling tube
1 piece of visking tubing
Mixture of 1% starch and 5% glucose solution
Iodine in potassium iodide solution
Benedict's solution
Bunsen and asbestos square
Bucket, marked 'Dirty Visking Tubing'
2 test tubes
1 test tube holder

Unit 5
None

Unit 6
1 beaker, 100 cm^3
1 test tube
1 test tube rack
1 test tube holder
Iodine in potassium iodide solution
1 glass rod
Beaker
Benedict's solution
Bunsen and asbestos square
1% starch suspension

Extension Work 1
Egg white suspension
1 test tube
Dilute hydrochloric acid
1 dropping pipette
1% pepsin
1 beaker
1% amylase

Extension Work 2
As for Extension Work 1
Also Universal Indicator Paper

Extension Work 3
None

Extension Work 4
1% starch suspension
1 piece of visking tubing
1 boiling tube
1 test tube rack
Benedict's solution
Bunsen and asbestos square